Production Planning and Scheduling in Flexible Assembly Systems

Springer
*Berlin
Heidelberg
New York
Barcelona
Hong Kong
London
Milan
Paris
Singapore
Tokyo*

Tadeusz Sawik

Production Planning and Scheduling in Flexible Assembly Systems

With 66 Figures
and 51 Tables

 Springer

Professor Dr. Tadeusz Sawik
University of Mining and Metallurgy
Faculty of Management, Department of Computer
Integrated Manufacturing
ul. Gramatyka 10
30-067 Krakow
Poland

ISBN 3-540-64998-0 Springer-Verlag Berlin Heidelberg New York

Library of Congress Cataloging-in-Publication Data
Die Deutsche Bibliothek – CIP-Einheitsaufnahme
Sawik, Tadeusz: Production planning and scheduling in flexible assembly systems: with 51 tables / Tadeusz Sawik. – Berlin; Heidelberg; New York; Barcelona; Hong Kong; London; Milan; Paris; Singapore; Tokyo: Springer, 1999
ISBN 3-540-64998-0

This work is subject to copyright. All rights are reserved, whether the whole or part of the material is concerned, specifically the rights of translation, reprinting, reuse of illustrations, recitation, broadcasting, reproduction on microfilm or in any other way, and storage in data banks. Duplication of this publication or parts thereof is permitted only under the provisions of the German Copyright Law of September 9, 1965, in its current version, and permission for use must always be obtained from Springer-Verlag. Violations are liable for prosecution under the German Copyright Law.

© Springer-Verlag Berlin · Heidelberg 1999
Printed in Germany

The use of general descriptive names, registered names, trademarks, etc. in this publication does not imply, even in the absence of a specific statement, that such names are exempt from the relevant protective laws and regulations and therefore free for general use.

Hardcover-Design: Erich Kirchner, Heidelberg

SPIN 10656007   42/2202-5 4 3 2 1 0 – Printed on acid-free paper

To my parents,

and Bartek

Preface

Flexible assembly systems (FASs) have emerged as a result of the developments in manufacturing and computer technology. Current market requirements characterized by

- increasing number of different types and versions of products,
- smaller batch sizes, and
- shorter life-time of products,

strongly determine the competitiveness in production assembly and additionally contribute to the development of flexible automated assembly. For example, at the end of 1986 [33] 40% of Japanese robots were specialized in assembly as compared with only 10% of European robots. The remaining 90% were used in welding, painting, and handling. The introduction of flexible automated assembly to high-tech sectors where assembly costs are critical is the aim of major European projects such as ESPRIT and BRITE programmes and the FAMOS-EUREKA project, e.g., [33, 34].

The book deals with production planning and scheduling in flexible assembly systems. The reader is familiarized with the FAS planning and scheduling issues for which various operations research modelling and solution approaches are discussed. In particular, applications of integer programming to the FAS short-term planning and fast combinatorial heuristics to the FAS scheduling are discussed.

The material in the book has been divided into seven chapters.

Chapter 1 presents the overall structure and hardware components and features of a flexible assembly system. The FASs classification is provided and illustrated with industrial applications of mechanical part assembly and printed circuit board (PCB) assembly.

Chapter 2 discusses major issues in the design, planning and scheduling of flexible assembly. Basic configurations of FASs and material flow networks are presented and various approaches to design for automated assembly and to assembly planning are discussed. The FAS production planning and scheduling are considered within a hierarchical framework with machine loading and assembly routing at an upper level and machine and vehicle scheduling at a lower level. Finally, specific issues in planning and scheduling of PCB assembly are discussed.

VIII   Preface

In Chapter 3 various bi-objective integer programming models and solution approaches are presented for machine loading and assembly-routing in FASs. An interactive procedure is proposed for simultaneous loading and routing based on weighting approach and a lexicographic algorithm is given for sequential loading and routing with a linear relaxation loading heuristic and a network flow routing model. Numerical examples illustrate possible applications of the modelling and solution approaches presented.

In Chapter 4 the sequential modelling and solution approach proposed in Chapter 3 has been extended for a bicriterion machine loading and assembly routing with simultaneous assembly plan selection in a general FAS and in a flexible assembly line. Numerical examples are provided to illustrate possible applications of the approach proposed.

Chapter 5 presents mathematical programming formulations for simultaneous loading and scheduling in flexible assembly cells. The formulations are illustrated with practical applications in mechanical part assembly with a robot assembly cell and in PCB assembly on a component placement machine.

Chapter 6 is devoted to production scheduling in flexible assembly lines where several assembly stages in series are either separated by finite intermediate buffers or there are no buffers between the stages, and each stage consists of one or more identical parallel machines. Fast push-type scheduling heuristics are proposed for the line with limited intermediate buffers or the line with no in-process buffers. For a comparison, a pull-type scheduling strategy is illustrated with some recent results for the Just-In-Time and multilevel scheduling of flexible assembly lines. Numerical examples provide the reader with possible applications of the various modelling and solution approaches presented.

In Chapter 7 simultaneous scheduling of assembly stations and automated guided vehicles is discussed for a general FAS and two different solution approaches are presented: (i) a multi-level approach, in which first machine loading and assembly routing problem is solved and then, given task assignments and assembly routes selected, detailed machine and vehicle schedules are determined; (ii) a single-level approach, in which machine and vehicle schedules are directly determined with no initial loading and routing decisions required. For each approach a scheduling algorithm based on dynamic complex dispatching rules is proposed and numerical examples are provided to illustrate and compare the two scheduling approaches.

The material presented in the book is illustrated with numerous examples, figures and extensive tables. The reader is provided with detailed mathematical models of the FAS planning and scheduling problems and descriptions of the solution algorithms proposed. Their applications are illustrated with many numerical examples and results of various computational experiments with the models and algorithms are reported.

The book is aimed primarily at students and professionals in production and operations management, industrial and systems engineering, and automated manufacturing.

This book benefited from numerous discussions with my colleagues. Professor Andreas Drexl and Dr. Rainer Kolisch from the Christian-Albrechts University of Kiel deserve special thanks for the careful reading of various parts of the manuscript and their valuable comments.

The book has been prepared with partial support by KBN research grant # 8 T11F 015 13, AGH grant # 10.200.10, and TEMPUS-PHARE project # S_ JEP-09434-95.

Tadeusz Sawik
Department of Computer Integrated Manufacturing
Faculty of Management
University of Mining and Metallurgy
Kraków, Poland

Table of Contents

1. **Flexible Assembly Systems – Hardware Components and Features** 1
   1.1 Basic components of a FAS 1
       1.1.1 Robots 1
       1.1.2 Peripheral equipment 4
   1.2 Classification of flexible assembly systems 5
   1.3 Examples of industrial installations 8
       1.3.1 Mechanical assembly 8
       1.3.2 Printed circuit board assembly 9

2. **Issues in Design, Planning and Scheduling of Flexible Assembly** 17
   2.1 FAS design issues 18
   2.2 Network design for material flow systems 22
   2.3 Design for assembly 27
   2.4 Assembly planning 30
   2.5 Planning and scheduling 32
       2.5.1 Machine loading and assembly routing 35
       2.5.2 Machine and vehicle scheduling 37
       2.5.3 Planning and scheduling in electronics assembly .. 38

3. **Loading and Routing Decisions in Flexible Assembly Systems** 41
   3.1 Description of a flexible assembly system 43
   3.2 Optimization of station workloads and product movements .. 44
   3.3 Design and balancing of flexible assembly lines 50
   3.4 Numerical examples 52
   3.5 Simultaneous loading and routing 56
       3.5.1 Problem formulations 56
       3.5.2 An interactive heuristic for loading and routing . 60
       3.5.3 Numerical examples 61
   3.6 Sequential loading and routing 68
       3.6.1 Problem formulations 68
       3.6.2 Lexicographic approach to loading and routing ... 71

XII   Table of Contents

        3.6.3 A linear relaxation-based heuristic for loading 73
        3.6.4 Numerical examples 77

4. Loading and Routing Decisions with Assembly Plan Selection 87
   4.1 Simultaneous loading, routing and assembly plan selection in a general flexible assembly system 89
        4.1.1 Problem formulation 90
        4.1.2 A two-level loading, routing and assembly plan selection 92
        4.1.3 Numerical examples 94
   4.2 Simultaneous loading, routing and assembly plan selection in a flexible assembly line 97
        4.2.1 Problem formulation 98
        4.2.2 A two-level loading, routing and assembly plan selection 100
        4.2.3 Numerical examples 102

5. Loading and Scheduling in Flexible Assembly Cells 107
   5.1 Loading and scheduling in a robot assembly cell 107
        5.1.1 Loading and scheduling constraints 109
        5.1.2 Objective functions 112
   5.2 Loading and sequencing in printed circuit board assembly ... 113
        5.2.1 Component retrieval problem 114

6. Production Scheduling in Flexible Assembly Lines 117
   6.1 Flexible assembly line with limited intermediate buffers 118
        6.1.1 Scheduling algorithm 121
        6.1.2 Numerical examples 124
   6.2 Flexible assembly line with no in-process buffers 130
        6.2.1 Scheduling algorithm 133
        6.2.2 Numerical examples 135
   6.3 Global lower bounds for flexible flow lines with unlimited buffers 139
        6.3.1 Numerical example 140
   6.4 Just-in-time scheduling of flexible assembly lines 142
        6.4.1 Final assembly scheduling 143
        6.4.2 Numerical example 147
        6.4.3 Balanced scheduling of a multilevel flexible assembly line 148
   6.5 Multilevel scheduling of flexible assembly lines with limited intermediate buffers 152
        6.5.1 Multilevel programming formulation 153
        6.5.2 Algorithm for multilevel scheduling 156
        6.5.3 Numerical example 159

7. Machine and Vehicle Scheduling in Flexible Assembly Systems 163
   7.1 Dispatching scheduling 165
       7.1.1 Dispatching scheduling of assembly operations 166
       7.1.2 Dispatching scheduling of transportation operations .. 166
   7.2 Machine and vehicle scheduling – a multi-level approach 167
       7.2.1 Problem variables 169
       7.2.2 Constraints 172
       7.2.3 Dispatching rules for machine and vehicle scheduling.. 174
       7.2.4 Scheduling algorithm 176
       7.2.5 Numerical examples 177
   7.3 Machine and vehicle scheduling – a single-level approach 181
       7.3.1 Dispatching rules for machine and vehicle scheduling.. 183
       7.3.2 Scheduling algorithm 185
       7.3.3 Numerical examples 188

References 197

Index 205

1. Flexible Assembly Systems – Hardware Components and Features

A *flexible assembly system* (FAS) is a fully integrated production system consisting of computer numerically controlled assembly stations, connected by an automated material handling system, all under the control of a central computer. A FAS is capable of simultaneously assemble a variety of product types in small to medium-sized batches and at high rate comparable to that of conventional transfer lines designed for high volume/low variety manufacture.

Flexible assembly systems have been first introduced in electronic and semiconductor industry, e.g., for printed circuit boards (PCBs) automated assembly. The progress in robot design observed over the past 20 years has made it possible to introduce flexible automation also in a more complicated electromechanical assembly.

The aim of this chapter is to present basic hardware components of flexible assembly systems and discuss their characteristics, classify the systems and describe some real world applications.

1.1 Basic components of a FAS

1.1.1 Robots

Robots constitute the most important components of a FAS. They perform basic assembly operations in electromechanical assembly. In [28] robots are categorized according to their functional and spatial features.

Functional features of assembly robots:

- number of degrees of freedom;
- speed;
- accuracy and repeatability;
- payload;
- working space;
- the possibility of including external sensors.

Spatial features of assembly robots:

- a fixed base: attached to the ground, a wall, or hanging from the ceiling;
- a mobile base: generally a translational axis;

– a gantry structure: where the robots hangs and can move along one or two translational axes.

The ideal assembly robot should have the following features:
– rigidity;
– minimal weight;
– quick response;
– maximal number of degrees of freedom;
– high speed;
– good repeatability;
– good accuracy.

Studies in electromechanical assemblies, described to [28], has led to formulating the following conlusions: in over 90% cases,
– the projection of the space required to assemble a product is limited to a surface smaller than 400 cm^2;
– parts moved weigh less than 2 kg;
– the assemblies contain fewer than 25 parts.

Moreover, half of the assemblies considered had only one assembly direction. In 20% of cases, there was a change in direction and, in 15% – two changes of direction. The assembly operations involved were: placing parts (40%), inserting (30%), and fastening (24%).

A survey (see, [28]) of 22 German industrial companies (car manufacturing, electromechanical, electric and electronic constructions, precision mechanics), where in total one hundred and sixty types of assembly tasks were examined has led to similar conclusions:
– 90% of the parts moved weigh less than 1 kg;
– 76% of products are assembled along three perpendicular axes. Those assembled in one direction make up 59% of all cases;
– 98% of products are made of fewer than 25 parts.

The first assembly robots were anthropomorphic: they were designed to replace an operator by copying the operator's movements. In order to meet the specific requirements of electromechanical assembly, a new type of robot was developed: the SCARA (Selective Compliance Arm for Robotic Assembly) robot. This robot was specially designed for vertical insertion of parts. It has only four degrees of freedom (two horizontal rotations, one vertical translation, and a horizontal rotation of the tool) – Fig. 1.1.

The SCARA robot is capable of high speed (about 10 m/s at the end of the arm) and accelerations. Its simplified kinematic chain leads to excellent repeatability and absolute accuracy of about 0.01 mm. Fig. 1.2 shows an application SCARA robots in electronics asssembly.

In order to increase productivity of robotized assembly, stations with two or more robots are designed. Cooperation between robots can be organized in several ways as described in [28].

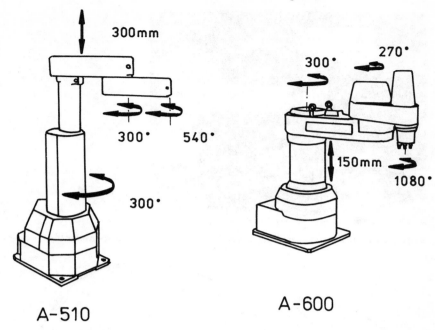

Fig. 1.1. The SCARA assembly robots FANUC A-510 and FANUC A-600

1. Each robot performs part of the task, but it is capable of performing some of the other robots task. This cooperation increases the system reliability; in case of a robot failure another robot can continue the assembly process. Such failures must be detected and another robot programmed to perform the tasks.
2. Each robot can work on a portion of the same part, which shortens the assembly cycle time.
3. Each robot can perform part of the total task but the working space may be shared among the robots. This organization of work reduces the space required, shortens the assembly cycle time, and requires less peripheral equipment (grippers, screwdrivers, etc.).
   Fig. 1.3 ([8]) shows a working space of two cooperating IBM 7545 (SCARA type) robots positioned so that they overlap one another. With the given robot's dimensions, a rectangular tool magazine of about 0.06 m^2 can be located in the overlapping area, and the distance between the robot is about 0.8 m.
4. Certain operations require total cooperation between two robots. For example, when a part is too cumbersome to be moved by a single robot.
5. Complementary capabilities of two different robots are required. For example, one may move quickly and have a good payload but its accuracy is poor. Such a robot is well suited for moving heavy parts. The other

4   1. Flexible Assembly Systems – Hardware Components and Features

Fig. 1.2. Application of robots FANUC A-510 in electronics assembly

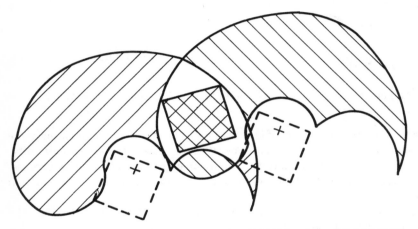

Fig. 1.3. Working space of two cooperating SCARA robots IBM 7545 [8]

robot has a good accuracy at the expense of speed and payload. It will be used for accurate insertion.

Cooperation between two or more robots improves the system flexibility, however it complicates control software.

1.1.2 Peripheral equipment

In [28] five types of complementary equipment required for assembly operations are distinguished.

1. Accessories required for the assembly operations themselves:

- Tools that execute a fastening process (pneumatic screwdrivers, presses, etc.).
- Grippers (magnetic, vacuum, adhesive, etc.) used to transfer and position parts or tools.
- Fixtures used to hold the components in place during construction of a subassembly.
2. Specific machines used for certain secondary operations, such as on-site production of parts with complicated shapes to avoid tangling during storage.
3. On-site storage devices such as indexed rotating tables or part feeders for standard elements.

   In mechanical assembly special-purpose or programmable part feeders are located at assembly stations to magazine and present to robots small parts used in large quantities such as screws, nuts, etc. The parts are usually presented in the correct orientation. Part feeders can be classified in three categories (see [15]):
   - *bowl feeders* that use vibratory action to move parts from the bottom of the bowl up an internal ramp, being positioned using gravity devices along the way;
   - *autoscrewdrivers* feed screws, nuts, rivets using an air/vacuum system to hold the part;
   - *gravity feeders* consisting of ramps or tubes which hold the part and are used for parts that are not conducive to vibratory feeding.
4. Material handling devices allowing entry into and exit from the assembly station such as belt conveyors or automated guided vehicles (AGVs).
5. Storage areas for subassemblies, faulty parts, and various tools.

1.2 Classification of flexible assembly systems

A basic component of various types flexible assembly systems is an *assembly station*. A robotized assembly station is made up of an assembly robot and various complementary equipment such as tool magazines, part feeders, pallet changers, input/output buffers, etc., all located in a finite work space of the robot (see Fig. 1.4).

In electronic and semiconductor industry an assembly station can be a special-purpose flexible machine such as automatic insertion machine for printed circuit boards assembly, see Sect. 1.3.2

The various types of flexible assembly system can be classified based on the system layout, material flow configuration, types of assembly machines used and its operational environment relating to high volume/low variety and low volume/high variety production.

1. **Flexible assembly cell** (Fig. 1.5) – represents a typical FAS configuration made up of one or more assembly stations connected by an au-

6   1. Flexible Assembly Systems – Hardware Components and Features

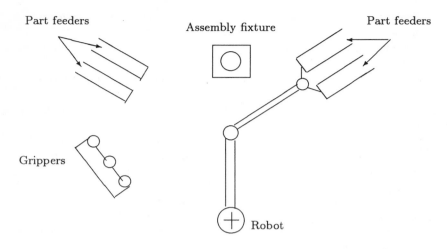

Fig. 1.4. A robot assembly station

tomated material handling system, with an intelligent control system capable of making decisions and managing the flow of materials and information. A robotized cell may include assembly stations with a single robot or two cooperating robots.

Fig. 1.5. Flexible assembly cell

2. **Flexible assembly line** (Fig. 1.6, 1.7) – is a unidirectional flow type system organized as a series of dedicated, special-purpose assembly stations linked with an automated material handling system. Each station

consists of a single machine or of identical parallel machines, usually assembly robots

Fig. 1.6. Flexible assembly line: mechanical assembly

Fig. 1.7. Flexible assembly line: electronics assembly

3. **General type flexible assembly system** may comprise a number of flexible assembly cells, assembly lines or single assembly stations connected by an automated material handling system, all under the control of a central computer.

The use of versatile, general-purpose machines in a flexible assembly cell results in high production flexibility, while dedicated special-purpose machines yield high productivity.

Flexible assembly line is used for high volume/low variety production of a few products that have stable designs and demand requirements. The line can be compared with a conventional transfer line, however it is capable of frequent and quick changeovers. The use of dedicated, special-purpose machines yields high productivity, and its capability of switching among operations allows for mixed-model operation.

A general type flexible assembly system is the most complex type of FASs usually used for low volume/high variety production. It can be compared with a job shop under conventional manufacture. The use of versatile, general-purpose machines results in high production flexibility. These systems usually produce to order, a large variety of products in varying, but small order quanities.

A typical assembly process in a FAS proceeds as follows: a base part of an assembly is loaded on a pallet and enters the FAS at the L/U (loading/unloading) station. As the pallet is carried by conveyors or automated guided vehicles through assembly stations, components are assembled with the base part. When all the required components are assembled with the base part, it is carried back to the L/U station and the complete assembly leaves the FAS.

1.3 Examples of industrial installations

In this section examples of the FAS installations in electromechanical and electronics industry are described. First, a multirobot assembly cell for mechanical assembly of an ignitor is presented, and then two types of component placement machines used in electronics industry for printed circuit board assembly are described.

1.3.1 Mechanical assembly

An ignitor assembly cell with two robots operating simultaneously is described in [83]. A cross-sectional diagram and disassembly of the ignitor is provided in Fig. 1.8 ([126]), and the assembly cell layout is presented in Fig. 1.9 ([83]), where the location of each machine are shown and the robot working spaces are indicated. The assembly cell consists of two robots (R1, R2), wire dispenser and cutter (WDC), powder dispenser (PD), press (P), ohm-test and inspection bin (OIB), part feeders and conveyors.

The assembly process proceeds as follows. The nylon housing and the steel cap are supplied into the WDC machine on separate conveyors. The conveyor next to the WDC machine brings in the steel body with the washer inside.

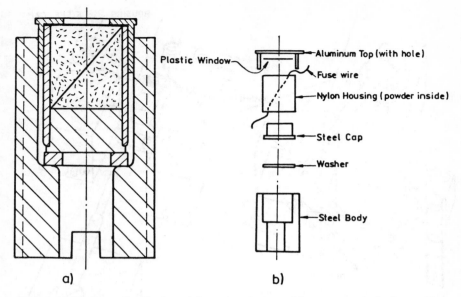

Fig. 1.8. Product description: (a) sectional view, (b) disassembly [83]

Robot R1 picks up the housing and rotates, wires (the fuse wire) and fits it on the cap, and then drops the capped and wired housing into the body. Next, it grabs the body and carries this subassembly over to the PD machine. Robot R1 either leaves the subassembly on a temporary storage area on the PD machine or holds it right below the powder-filling chute of the PD machine and attends the powder filling. If it does the former, then robot R2 can pick up the subassembly and subsequently attend the powder filling.

Once the powder filling is done, the subassembly is moved to the P machine by either robot. At the P machine, the subassembly is either left in a temporary storage on the P by the attending robot, or it is carried on, by any robot, to be held in front of the window and the top carriers to collect these two final pieces. Then the assembly gets pressed by the P machine. If it is left unpressed in the storage area by one robot, then the other robot can pick it up and attend the pressing operation. After the assembly is pressed, the complete unit is moved to the OIB by robot R2 only, since robot R1 cannot reach the OIB. Based on the inspection results, robot R2 passes the ignitor assemblies into two separate outgoing troughs.

1.3.2 Printed circuit board assembly

Assembly of printed circuit boards is one of the most important assembly process in electronics industry, where flexible automation is widely used. In recent years Surface Mount Technology (SMT) has begun to replace the standard

10   1. Flexible Assembly Systems – Hardware Components and Features

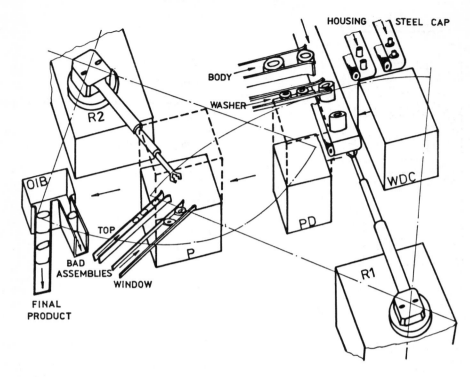

Fig. 1.9. A multirobot assembly cell [83]

pin-through-hole design of electronic components. SMT technology permits the leads to sit on the surface, and there is no need for complex finger manipulations to position the leads as in pin-through-hole insertion. However, component placement must be increasingly accurate due to smaller electrical connection pads and denser designs. No leads extending through the board enable the components to be placed on either side with proportionate benefits. The large number of components placed on a single board and the precision required make it necessary to use highly automated machines for PCB assembly, called component placement machines.

In this subsection two types of component placement machines are described:

– Dynapert MPS 500 placement machine;
– Fuji CP II placement machine.

The Dynapert MPS 500 placement machine. Fig. 1.10 illustrates a Dynapert MPS 500, a computer numerically controlled dual delivery pick-and-place machine (see [3, 5]) for populating printed circuit boards with Surface Mount Technology.

1.3 Examples of industrial installations   11

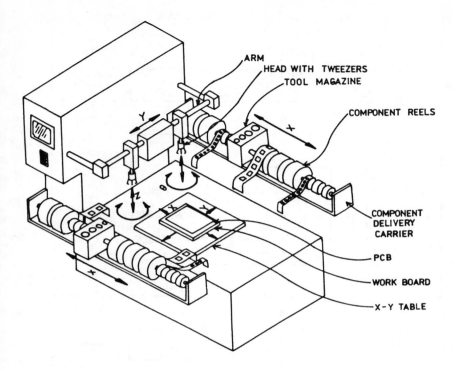

Fig. 1.10. Dynapert MPS 500, dual delivery pick-and-place machine [5]

The key components of the Dynapert MPS 500 machine include the work board, the placement arm, two pick-and-place heads, and the two component delivery carriers. These components are capable of independent movements as shown in Fig. 1.10. Exploitation of these concurrent movements requires coordination of the movements. Fig. 1.11, reproduced from [3] shows a schematic description of the main features of the pick-and-place operation. The pick-and-place heads are mounted on the two ends of a fixed length arm which can move between two fixed position in the y-direction only. The heads are capable of making a vacuum pickup, rotating, tweezing and making a reverse vacuum placement. (Tweezing is an operation to center and position a component after it is picked up by the vacuum nozzle.) The two component delivery carriers move in the x-direction only. Each carrier has a tool magazine that can hold four pickup-and-placement nozzles (tools) and sixty slots for accommodating reels of components. Component pickup by the head requires the carrier should move and position the slot with the required components in the fixed pick position. The workboard can move in both the x and y directions and should be aligned under the head to perform the placement operation.

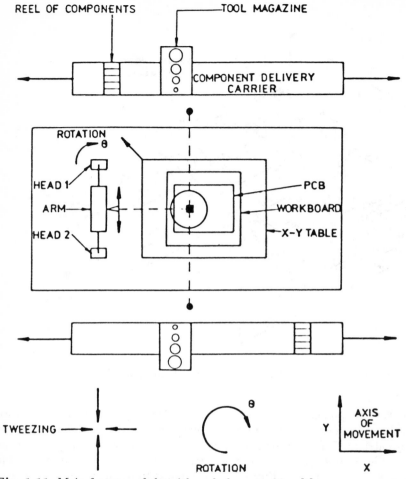

Fig. 1.11. Main features of the pick-and-place machine. [5]

Typically, a pick-and-place operation requires the following steps, described in [5]:

1– movement of the arm to place the head in the pick position;
2– movement of the carrier to position the component reel in the pick position;
3– pick operation by the head, rotation and tweezing of the component;
4– movement of the arm to locate the head in the place position;
5– movement of the board to align for placement operation;
6– placement operation.

Steps 1 – 6 are repeated for each pick-and-place operation alternating between the two sides. The above description assumes that no nozzle change is required. Nozzle changes are usually necessary when components belong to different groups. In that case, the tool magazine needs to be in the pick position to facilitate nozzle changes. Nozzle changes represent unproductive movements and tend to reduce the production rate ([5]).

To realize the full benefits of the machine, it is necessary to synchronize the concurrent operations of the two carriers and the head such that, while one head is performing the pick operation, the other is engaged in the placement operation. The production rate can be improved by optimization of the placement sequencing problem along with the assignment of reels of components to slots on the carriers [5], see Sect. 2.5.2

Fuji CP II placement machine. The Fuji CP II is one of the most advanced component placement machines. It has a placement rate of over 12 000 components per hour with an error rate less than 1 in 10^4, and it is equipped with a stereoscopic vision system for inspection (cf. [20]). The main parts of Fuji CP II include the PCB table, the magazine rack, and the placement head carousel, see Fig. 1.12 [20, 54]. During the assembly the board lies on a movable table in the center of the machine with two axes of motion. For each component to be mounted, the table moves the board so that it is directly under the fixed placement station of the carousel. A magazine rack contains a number of slots to which feeder tapes can be assigned. The same component type can be assigned to more than one different slots. The rack can move back and forth along a single axis so that a reel with the required component type is brought in line with the pickup station. Above the PCB table is a carousel containing 12 heads, and it can simultaneously hold up to six components. Each head has a pair of nozzles of different diameter to grip and hold the components until their placement on the PCB. Only one nozzle is used at a time. The carousel rotates clockwise so that components gripped from a slot of the magazine rack are next mounted on the board. However, as the carousel rotates, several additional operations are performed between the pickup and placement stations. An automated visual inspection system is used for each component examination for orientation and diagnostics. If a problem is identified, the component is ejected and reassigned to the end of the placement sequence. Otherwise it is oriented for placement. If any problems occur at the placement stage, a second ejection station is provided. The last two stations are used to set up and reorient the nozzle to pickup the next component. The additional operations are performed in parallel with the dominating gripping and placement operations and do not constrain the assembly schedule.

The table may start moving as soon as the previous component is placed. Similarly, the magazine movement may begin as soon as the previous gripping operation is completed. However, the carousel rotation may not start until both the placement and gripping operations are finished.

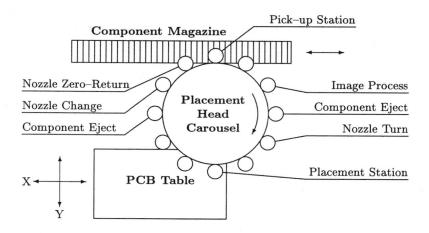

Fig. 1.12. The Fuji CP II [13]

The PCB table moves between two consecutive placement operations from the position at which the first component was placed to the position required to place the second component and the carousel rotates 30° clockwise so as to position the head right above the placement location and the diametrically opposed head at the pickup station. Similarly, between two consecutive gripping operations, the magazine rack shifts to retrieve the required component.

Briefly, the assembly process proceeds as follows. Assume that the placement of the ith component on the board has just been completed and the table starts to move to the $(i+1)$st placement location. The pickup head is ready to retrieve the $(i+6)$th component from the magazine rack as soon as the appropriate slot has been been positioned at the pickup station. Once this is done, the head proceeds with gripping the $(i+6)$th component from this feeder slot. After the gripping operation has been completed, the magazine rack starts to shift to position the slot with the $(i+7)$th component at the pickup station. The carousel rotation begins only after the ith component has been placed on the board and the $(i+1)$st component has been retrieved from the magazine. The carousel rotates 30° clockwise to prepare the placement of the $(i+1)$st component and the retrieving of the $(i+7)$th component.

The duration of rack movements depends on the distance between slots with the component types selected for consecutive gripping operations. Therefore, given the component placement sequence and the assignment of component types to slots of the magazine, the assembly makespan depends on the selection of feeder slots from which the components are retrieved if more than one slot is assigned the same component type, e.g., [20, 26, 54].

It is important to observe that placement operation i and gripping operation $(i+6)$th do not need to be performed simultaneously, however they must be carried out between two consecutive rotations of the carousel. The table and the magazine rack may move concurrently with each other, and with the carousel rotation.

The above properties do not hold for the Fuji CP IV placement machine (e.g., [54]) where various placing techniques (insertion, onsertion, glueing) are used. For the Fuji CP IV, the placement of component i and the gripping of component $(i+6)$ starts and ends at the same time. This implies that as soon as the placement and the gripping operations are completed, the three devices: the table, the magazine rack and the carousel start to move at the same time, which considerably simplifies control of the assembly process. The movement that takes the maximum time determines the time at which a new iteration starts.

2. Issues in Design, Planning and Scheduling of Flexible Assembly

A flexible assembly system is an extremely complex, large-scale system consisting of many interconnected components of hardware and software. An enormous volume of data is required to describe and manage it. In order to reduce the system complexity in its describing, planning and managing, the global objective of best utilization the FAS capablities can be transformed into various local objectives based on hierarchical decomposition.

The various issues in the FAS design, planning, scheduling and control can be considered within a hierarchical framework shown in Fig. 2.1 comprising three interconnected decision levels (cf. [114, 119]):

- FAS design (long-term).
- FAS planning (medium-term).
- FAS scheduling and control (short-term).

Decisions at each level differ in scope, detail and time horizon addressed.

FAS design problems include the selection and layout of assembly machines and material handling system, design of products for automated assembly and planning of assembly. These decisions typically have long-term implications and are consequently made only periodically.

FAS planning problems address resource allocation in the medium-term and include machine loading and assembly routing. Simultaneously assembly plans for products can also be selected.

FAS *scheduling and control* problems relate to execution of production orders in the short-term and include base part input sequencing, machine and vehicle scheduling, monitoring system performance and taking the necessary corrective actions.

Fig. 2.1 indicates the different information situation and the different objectives of the different decision levels. The management hierarchy in Fig. 2.1 can be considered a typical production planning and control hierarchy with the budgeting level as the FAS design level and with two operative levels.

Design level decisions are driven by strategic decisions that include issues such as justification of FASs and long-term competitive advantage. The resulting design, in turn, will affect planning and scheduling decisions.

The various FAS design, planning and scheduling problems are presented in the following sections.

18   2. Issues in Design, Planning and Scheduling of Flexible Assembly

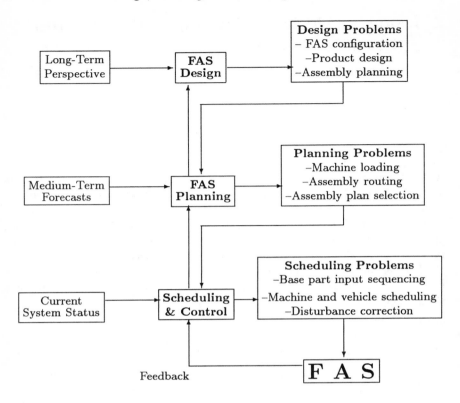

Fig. 2.1. Production planning and scheduling in a FAS

2.1 FAS design issues

A typical design process for assembly system can be described in three stages (see, [64]). The first stage determines and characterizes three key components of the system: products assembled, machines used, and material handling system used. For each component, designers usually have many alternatives, each with different features and costs. Once the alternatives of the three components are decided and characterized, the second stage integrates them and generates design alternatives by determining capacities of various resources, layout and other policies required to operate the system. The third stage evaluates these design alternatives to see if they are economically justified and can produce required production volumes with good quality standard. The design process with many alternatives is usually repeated before deciding on one to be installed.

A typical FAS consists of a set of assembly stations and a load/unload (L/U) station connected by conveyors or transporter paths. The flexible assembly machines (e.g., assembly robots or automatic insertion machines) have a finite work space due to their physical configuration. Because a component feeding mechanism associated with each assembly task uses some of the finite work space, only a limited number of tasks can be assigned to a robot. The finite number of tasks is called the flexibility capacity (e.g., [64]). There are negligible setup times between task changes among the tasks assigned to a robot. Components and assembly tools are always available when a base part is ready to be assembled at each station. The assembly tools are less perishable than cutting tools in flexible machining, and computer controllers at each station keep track of component delivery, e.g., [63, 64].

The assembly times are relatively small and comparable with vehicle transfer times. Therefore, material handling system in a FAS should have sufficient excess capacity to avoid bottlenecks in the system and its underutilization. The FAS designer usually seeks the minimum number of pallets required to satisfy production requirements. Keeping the number of pallets/fixtures at the minimum ensures not only a small amount of material handling equipment and small work-in-process inventories, and consequently, short conveyor length, short transfer time, and small floor space. This is because the total number of pallets/fixtures circulating in the system usually remains unchanged throughout the entire production, see [64].

The FAS design process includes an evaluation of limited buffer spaces on performance of the system and seeks the smallest number of buffer spaces that satisfy production requirements. Unnecessarily large buffer spaces cause wasted floor space and long travel times. On the other hand, too small buffer spaces cause machine blocking, resulting in low machine utilization.

The two important characteristics of a FAS can be described as follows (cf. [48]):

– Greater product customization, i.e., assembly to order at a relatively low unit cost.
– Dynamic reconfiguration of assembly systems to accommodate swift changes in the product design.

The system with such characteristics is capable of assembling a variety of high quality products at low cost. This requires that the FAS be simple and flexible and production planning and scheduling problems be solvable in a short time.

The above requirements are associated with the concept of agility, i.e., the ability of a production system to produce a variety of products of high quality at low cost (e.g., [48]). The agility concept has an impact on design of products and assembly systems. The complexity of the design problems can be reduced if enough consideration is given to concurrent design of products and systems, e.g., [61]. A product should be designed for an assembly system, as much as the assembly system should be designed for the product to simplify the planning and scheduling problems solving process and improve the system performance.

20   2. Issues in Design, Planning and Scheduling of Flexible Assembly

The following design rules can support design of products and assembly systems to meet the requirements of agile assembly ([48]):

1. *Design products with robust scheduling characteristics.*
   Product design has an impact on efficient scheduling and performance of an assembly system. The latter is due to the difficulty of rescheduling because of changes in production demand and product mix. Product design and its assembly plan should be robust and allow for easy reconfiguration of a FAS. For example, an assembly plan with a linear assembly structure (i.e., a chain of assembly tasks) simplifies scheduling irrespective to the changes in the product mix and demand.
2. *Design products to simplify the product flow in a FAS.*
   The flow of products is an important factor to be considered. In [48] four different product flows are discussed: *repeat operation, serial flow, by-pass flow, and backtracking flow* In addition, *the branch/merge flow* can be observed (Fig. 2.2 a–e).

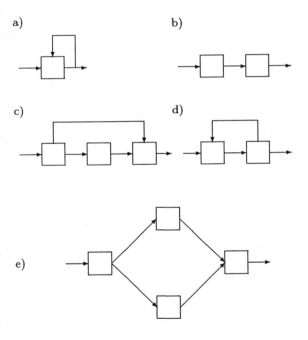

Fig. 2.2. Product flow types: a) repeat operation, b) serial flow, c) by-pass flow, d) backtracking flow, e) branch/merge flow [48]

Of this five flow types, the serial flow is the most desirable because it eases control of the assembly process and material handling. Backtracking is the least desirable flow since it complicates the flow. Backtracking flows can be eliminated by assigning operations forming a cycle to the same station or by redesigning the products so that operations forming the cycles can be combined into subassemblies.

3. *Reduce the number of assembly stages in a FAS.*
   The FAS configuration impacts on the complexity of the scheduling problem. For example, by reducing the length of an assembly line, the complexity of the scheduling problem can be reduced. It is well known from the flowshop scheduling theory that when the number of machines increases to 3, the scheduling problem becomes $\mathcal{NP} - hard$. In addition to high computational complexity, long lines are undesirable in terms of the line balancing efficiency and its reliability. The number of assembly stations should be minimized whenever possible. For example, different operations can be assigned to one station and multiple robots can be used. Fig. 2.3 (e.g., [48]) shows two different configurations of an assembly line.

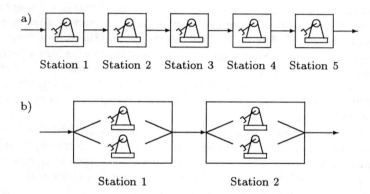

Fig. 2.3. The two different configurations of an assembly line: a) long line, b) short line [48]

4. *Streamline the flow of products in a FAS.*
   The configuration of a FAS determines the traffic and direction of product flow. The complexity of machine and vehicle scheduling increases with the number of assembly routes available and the number of direction flows. Since a job shop scheduling is usually more complex than the flowshop scheduling, the complexity of scheduling can be reduced if flow of products can be streamlined.

2.2 Network design for material flow systems

The effectiveness of a flexible assembly system depends on material handling system, and in particular on flow path design. The flow path design has a significant effect on the transportation time, the operating expenses and the installation costs of the material handling system. There is a strong relationship between the material handling design and the facility layout design (i.e., the location of assembly stations, loading/unloading stations, central buffers, etc.) [117]. The relationship between these two design functions can be expressed by the distance measure from pick-up station P to delivery station D used in the facility layout design, see [60]. In many facility layout design procedures a direct rectilinear distance measure from pick-up station to delivery station is used. In some cases, like monorails and conveyors direct rectilinear or Euclidean distance measures are usable. However, when AGVs are applied the intermachine flow is possible only via the aisle network.

Material handling systems are designed and analysed using various analytical approaches. The most common tools are mathematical programming, queueing networks models, graph theory algorithms, and simulation models. In order to determine flow paths, flow directions, and the station location, the mathematical programming and graph theory approaches are widely used. The material handling system is modelled as a network or a non-directed graph, where pick-up stations, delivery stations, and the intersections are represented as nodes, and the possible flow path segments are represented as edges of the graph.

On the other hand, to determine input/output buffers sizes, to calculate the number of AGVs, and to estimate the system throughput, queueing and simulation models are preferred. An important issue in designing of an AGV system, which is influenced by the flow paths topology and the system dynamics, is the number of carriers to be determined. An AGV can be in different states: travelling loaded to unload, travelling empty to load, idle, either travelling or stationary, blocked owing to congestion, loading or unloading, charging, etc. In order to determine the number of AGVs required one needs to estimate all these states (see, [117]). For example, empty carrier flows have a large impact on the network design and the number of AGVs required. An empty (deadheading) trip can be estimated as a portion of each transportation task. The empty trips starts from the point where the AGV receives the pick-up request. This point can be a delivery station, where the carrier just completed its last assignment, a random point where the idle carrier travelling around received the call, a fixed point where the idle carrier is parked waiting for an assignment. The empty trip ends at the pick-up station that issued the request.

Also, the blocking factor needs to be estimated. Blocking can occur owing to congestion, heavy traffic at intersections, communication delays, blocked pick-up and delivery stations, etc. The most common approach is to use an estimated proportion of carriers trips (loaded and unloaded) as the blocking time [117].

Classification of material flow networks. The various material flow configurations can be classified based on network topology, number of lanes and flow direction (see, [117]).

Network Topology
1. Conventional network.
2. Loop network.
3. Tree/spine network (e.g., Fig. 2.4 [117]).

Number of Lanes
1. Single lane: traffic is handled on one lane only.
2. Multiple lane: one lane can be reserved for slow loaded carriers, while the other lane is restricted to empty faster carriers.

Flow Direction
1. Unidirectional flow: the flow on each lane or set of lanes is fixed in one direction.
2. Bidirectional flow: the flow on all or some of the lanes is possible in both directions. Several options are possible. For example, one or more lanes can be dedicated for the flow in each direction or a single lane capable of switching flow direction is applied.

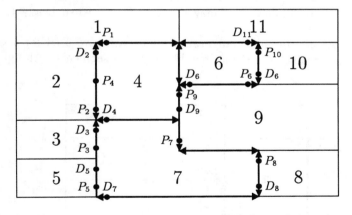

Fig. 2.4. Bidirectional single lane tree flow path [117]

Conventional network can be either single lane unidirectional or bidirectional. Three basic bidirectional conventional configurations are listed in [117]:

24   2. Issues in Design, Planning and Scheduling of Flexible Assembly

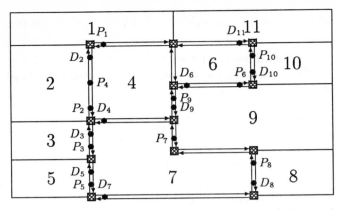

Fig. 2.5. Bidirectional conventional parallel (dual path) flow path [117]

1. *Parallel paths.* An example of this configuration shown in Fig. 2.5 ([117]) is comprised of two parallel unidirectional systems. It is the easiest bidirectional system to control, however a double set of tracks is required.
2. *Switchable paths.* In the configuration shown in Fig. 2.6 ([117]) a single track is used for flows in both directions. An AGV entering a lane segment determines the flow direction of that lane for the entire travel duration. Carriers wishing to travel in the same direction can do so and the rest wait until the lane is cleared. The configuration is complicated to control owing to flow conflicts and may not be efficient when heavy traffic is involved.
3. *Mixed systems.* In this configuration dual paths are used for segments with heavy traffic, and the switchable paths for segments with light traffic.

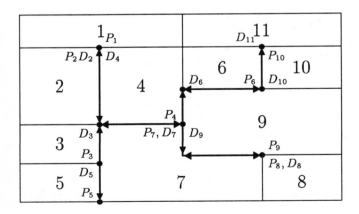

Fig. 2.6. Bidirectional conventional switchable (single path) flow path [117]

The unidirectional and bidirectional flows that are possible in the conventional configurations as well as shortcuts and alternative routes available contribute to the complexity of the control problems associated with the conventional systems. Simplification of the physical network and by this reduction of the number of conflict points may help to eliminate some of the decisions the controller has to make.

Single-loop flow path associated mainly with unidirectional flows, which contains no shortcuts and no alternative routes is another simplified configuration. In the case where the number of carriers is restricted to one per single loop it is possible to use bidirectional flows. The carrier can travel clockwise or counterclockwise, whichever is shorter. The advantages of the single-loop configuration are as follows ([117]):

- The design and control of the single-loop system is simplified while the performance offered is comparable to more complex systems. There are no intersections in the flow path, and hence collisions are more unlikely and the vehicle scheduling problems becomes much more simple. Lack of alternative routes in the flow path simplifies the routing decisions.
- The impact on the system performance of the dispatching rules selected for vehicle scheduling as well as of empty trips is reduced, owing to a similar performance across all rules.

The drawbacks of using a single-loop flow path configuration are as follows ([117]):

- The single-loop is less flexible. In the case of a failure somewhere along the loop, the entire system may be closed.
- No shortcuts available in the system requires extra transport capacity to be available. Once a station is passed the carrier needs to travel the entire length of the loop to get to the destination station again. The use of multi-load carriers can overcome this flaw with no additional congestion caused by more carriers in the system.
- The system throughput based on the single-loop flow path may be lower compared to a system based on a conventional flow path.

Some of this drawbacks can be alleviated through the use of bidirectional flows. However, implementation of bidirectional flow in a single-loop network can be difficult. An alternative approach is to use the *segmented bidirectional single-loop* (SBSL), see Fig. 2.7 [117]. The SBSL system consists of a single-loop flow path which is divided into non-overlapping single carrier segments. Transfer buffers are located at both ends of each segment and serve as input/output buffers. A carrier can deposit loads which are headed to other segments and pick-up loads from the other segments. The carrier has the capability to travel clockwise or counterclockwise on each segment, whichever is the shorter distance to its destination. The duration of the load/unload operations is usually in the range 15–30 s ([117]).

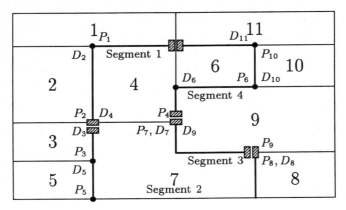

Fig. 2.7. The SBSL flow path topology with four segments [117]

The performance of the single-loop configuration can be improved by simply breaking up the loop into non-overlapping smaller loops so as to balance the workload of all zones. The resulting network is called a *tandem flow path configuration* (Fig. 2.8 [117]). The transfer between zones is done by local input/output buffers. Each loop is usually served by a single carrier that is allowed to travel in the clockwise or counterclockwise direction. The size of the zone is limited by the workload capacity of one carrier. The more loops the system contains the shorter the total travel time becomes. However, the waiting time at the transfer buffers increases.

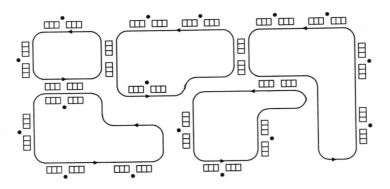

Fig. 2.8. Tandem flow path configuration [117]

The tree flow networks (Fig. 2.4 [117]) are common configurations for several material handling systems like cranes, monorails and conveyors. This is due to its simple and efficient structure. However, no shortcuts or alternative routes exist in this type of system. Using bidirectional flows may cause congestion, and hence the number of carriers should be limited, similarly to the bidirectional single-loop network.

In many production systems, flow requirements exist only between a few points as defined by a given assembly plan. Therefore, disjoint physical flows can support the logical material flow requirement, i.e., the flow graph can be a disjoint graph. In this type of system not every node can be reached from any other node. This type of flow can be supported by segmented flow topology (SFT). The *segmented flow topology* (SFT) is comprised of one or more zones, each of which is separated into non-overlapping segments with each segment serviced by a single carrier. The carrier can travel forward and backward on each segment whichever direction is a shorter distance to its destination. Transfer buffers are located at both ends of each segment (similar to the SBSL type network) and they serve as the interface between the segments. These buffers serve as input/output buffers where a carrier can deposit loads which are headed to other segments and pick-up loads from other segment.

Using the SFT means the shortest path from any pick-up station to any delivery station can be traversed. All segments are bidirectional and mutually exclusive so that no congestion and blocking are present.

The SFT design procedure is comprised of the following 5 steps ([117]):

1. Determine all the shortest path alternatives for each pair of pick-up and delivery points.
2. Determine location of P/D stations that minimizes the total cost of transportation and installation of additional P/D stations.
3. Determine the flow path network using the P/D station locations and the shortest path algorithm
4. Split the network into segments based on the number of carriers required for each zone
5. Calculate total cost of the material handling system.

2.3 Design for assembly

The assembly automation requires that the product be designed to make it easier to automate its manufacture. Robot lack such sophisticated tools like human hands, which can assemble in several directions or can hold parts in place while inserting other components. The product must therefore be redesigned to allow for automation of its assembly so that the individual operations become sufficiently simple for a machine to perform. In design of products and parts for automated assembly the following design rules have to be considered ([60]):

– Minimize the number of parts in a product.
– Ensure that the product has a suitable base part on which the assembly can be built.

28   2. Issues in Design, Planning and Scheduling of Flexible Assembly

- Ensure that the product base has features that will enable it to be readily located in a stable position in the horizontal plane.
- Design the product so that it can be assembled in layers.
- Design parts for self-alignment so that they fit together in only one way and do not require a secondary operation for alignment.
- Design parts to be compatible with grippers.
- Avoid expensive and time-consuming fastening operations such as screwing and soldering.
- Design parts to prevent jamming in the part feeders.
- Design a symmetrical parts to reduce handling and the need for sensors to detect features.
- If a part must be asymmetrical increase its asymmetry to eliminate jamming and misalignment during assembly.

In [28] the design for assembly techniques are classified into the following five groups :

1. **Specific assembly operation techniques.**
   This approach takes a microscopic view of a particular design rule and its application to the suitability for assembly of adjoining parts. In addition, it involves the analysis of component geometry and the mathematical conversion of the shape of a part into a value of assemblability to determine whether a specific operation is applicable. This approach has been applied only to insertion and tangling. It may also be useful when applied as part of a much broader methodology.
2. **Axiomatic approaches.**
   Axiomatic approaches view design as a mapping process from the functional domain to the physical domain, to the process domain. The first step of such an approach requires a definition of functional requirements, design parameters, and constraints. Axiomatic approaches have shown that design possesses two fundamental axioms:
   - maintaining the independence of functional requirements;
   - minimizing the information contents.
   Axiomatic approaches provide only broad rules for designers and hence are severely limited.
3. **Unstructured rules and concepts.**
   The majority of existing techniques for design for assembly falls into this category. Literature presents collections of design rules and concepts which are derived from product structures and features, providing good inherent assembly characteristics. However, reliance on unstructured rules and guidelines has many disadvantages. The rules and quidelines are too general, may conflict and require the large number of considerations to be taken into account by the designer (see [28]).
4. **Procedural methods.**
   The procedural application of rules provides systematic procedures as detailed in written checklists or computer software. Once a product is

designed, the assemblability factors for each component part are gathered and assesed. This is usually done using either spreadsheet analysis or rule-based systems. The procedural methods give little advice on how to improve the design and the designer can only go through the procedure from top to bottom (i.e., from assembly method selection to calculating design efficiency).

5. **Expert/knowledge-based systems.**
Recent efforts have focused on developing artificial intelligence-based approaches to design for assembly. However, they can be used in only very small application area.

One of the most well-known method of design for assembly is a set of procedures developed by Boothroyd and Dewhurst at the University of Massachusetts and later at the University of Rhode Island ([23, 24]). The method consists of two parts ([15]). The first part is a catalogue of generic part shapes and types classified by group technology methods according to ease of feeding by part feeders and ease of assembly by manual or automatic means. Estimates are given for assembly times. Examples are parts that are assembled by pushing, pushing and twisting, pushing, twisting and tilting, etc. A designer can estimate the time of assembly for all parts by consulting this catalogue. Assembly times usually are directly proportional to assembly costs.

The second part is a source of rules, advice or prompting questions concerning good design for assembly practice to help the designer achieve some savings in assembly costs. The designer is provided with a set of software modules to support the design process. The user begins by knowing the design of each of the assembled components and an exploded view of the assembly. The first package assist the designer in determining if the product is a candidate for automatic assembly. The software queries the user as to the values of various parameters that are important in automated assembly. Example parameters include number of parts in the assembly, annual production per shift, capital expenditure, total number of parts for building different product styles, and annual cost of one assembly operator. The results are presented as a list of costs for using several assembly systems together with estimated costs of manual or automated assembly versus production volume.

Once it is determined which of the assembly methods will be suitable, the user enters a procedure to optimize the assembly itself. Part codes are given based on answers to questions about geometry, function, and foreseen problems such as tangling or nesting during feeding. The codes give the designer an idea of assembly costs and possible locations of bottlenecks in assembly. Finally, the designer is asked questions to determine the theoretical minimum number of parts in the assembly based on relative part motion, different required materials, and repeated assembly-disassembly of certain parts to allow assembly of other parts. This would eliminate parts and will reduce assembly cost.

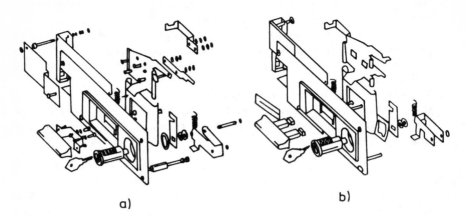

Fig. 2.9. Xerox latch mechanism design [15]

As an example, consider the latch mechanism in Fig. 2.9 ([15]), which shows an existing design (a) and a proposed design (b) for ease of assembly ([29]). The new design eliminates 73% of the parts, reduces assembly time by 79%, reduces assembly cost by 79%, and reduces parts cost by 24% for an overall savings of 36% in total product cost. This overall savings was accomplished by removing fasteners (note the lack of screws and nuts in the improved design), redesigning components to be multifunctional, and changing part materials.

2.4 Assembly planning

Assembly planning generates plans based on five types of information (see, [28]):

- geometrical (dimensions of components and their relative positions in the final assembly);
- components attributes (the components physical characteristics and the standard categories of parts);
- final assembly information (assembly directions);
- topological (relations between parts of the assembly);
- technological (additional constraints due to a prior knowledge of assembly devices).

An *assembly plan* is a graph of precedence relations in which nodes represent assembly operations and arcs represent precedence relations. In general, the plan is independent on the production means to be used, however knowledge of the assembly system makes it possible to choose among the various

2.4 Assembly planning

plans available. The data elements composing these plans are the assembly actions or operations, in the form of object-level instructions such as:

− insert peg a in hole b;
− place part c on part d;
− screw part e onto part f.

These are the operations required to construct the product. They can be performed serially and/or in parallel, and hence the plan usually lists a set of sequential actions called an *assembly sequence*.

An optimal assembly sequence is such that incurres the minimum assembly costs. The minimum cost means the minimum number of assembly operations required, the minimum number reorientations of the parts and the most reliable sequence. One can think of a set of parts as having assembly liaisons, that is, relative motion and connections between two mating parts. As an example consider the rear axle assembly shown in Fig. 2.10 ([15, 126]).

The parts and their assembly liaisons indicating how the various components are related are listed below (cf. [15]).

Parts	Assembly liaisons
A − carries assembly	1 − C to A
B − backing plate with brake shoes	2 − B to A
C − axle shaft	3 − J to B
D − brake drum and nut	4 − D to C
E − withdrawn gear shaft and bolt	5 − G to C
F − inserted gear shaft and bolt	6 − E to A
G − push in shaft and C washer and push shaft out	7 − F to A
H − oil	8 − L to A
I − cover	9 − I to A
J − brake cable, coiled	10 − H to A
K − final pressure test	11 − K to A
L − air test plug	12 − J to C

The above list does not show the relative precedence relationships among the components. For example, liaisons 2 and either 5 or 6 should be done before liaison 8. That is, the brake shoe backing plates and the gear shaft must be assembled to the carrier assembly before the oil can be filled in the housing. Otherwise, the oil would leak out.

Precedence constraints such as these can be written down and used to determine all feasible assembly sequences allowed. A graph of the possible assembly processes can be seen in Fig. 2.11 ([15, 126]). The 12-element matrix represents the 12 liaisons for the assembly, and the black filled squares indicate that a particular liaison has been accomplished. Note that all possible assembly sequences are shown as all possible vertical paths in the graph. There are, in fact, 330 different sequences possible. The problem is to come up with the best assembly sequence, e.g., based on the least cost. Reducing the number of sequences would make this analysis a bit less complex.

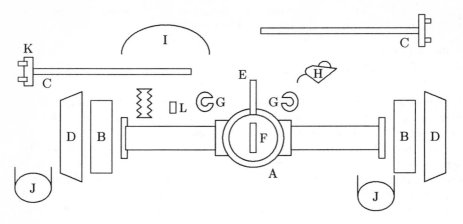

Fig. 2.10. Parts of rear axle [15]

One can reduce the total number of paths in the graph by making some simplifying assumptions about the assembly process. For example:

1. The axles are to be fastened with C washers following axle-shaft insertion to avoid some unstable assembly states and reduce the risk of the assembly's coming apart.
2. Wrap and attach the brake cables as soon as possible to prevent damage to the cables.
3. Place the cover on immediately after filling the carrier with oil to prevent oil spillage.
4. Insert the plug immediately after the leak check to avoid leakage.

By applying these four common-sense restrictions on the process, the assembly graph becomes as shown in Fig. 2.12. Note that now there are only 6 different assembly paths out of 330 original sequences. This technique of establishing liaisons and using graph to show possible assembly steps can help organize an assembly.

Another method for establishing an assembly sequence is to determine all possible disassembly sequences and then reverse them. This technique is sometimes easier to conceptualize than the assembly process because of its constrained nature.

2.5 Planning and scheduling

The purpose of the *FAS planning* is to generate a pre-production system setup to prepare its resources for completing the production order during the upcoming time horizon, whereas the *FAS scheduling* deals with the current

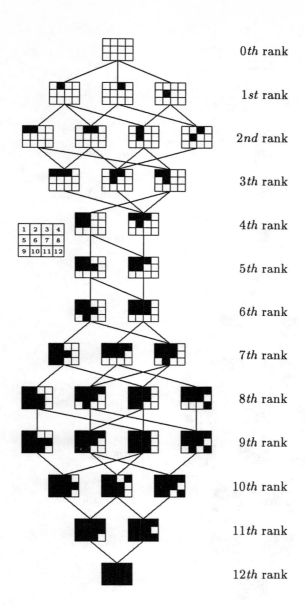

Fig. 2.11. Rear axle assembly sequences [15]

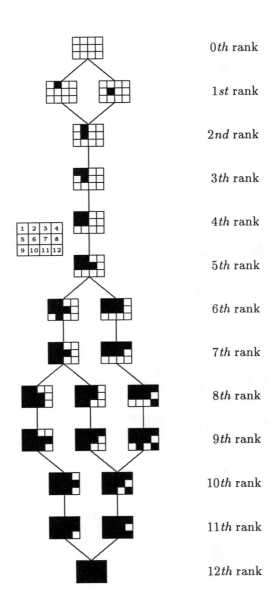

Fig. 2.12. Remaining rear axle assembly sequences after imposing restrictions [15]

control of product flows through the system during actual execution of the order.

The main issue in FAS planning is to determine an allocation of the FAS resources among the production tasks so as to complete the tasks and to optimize some system performance measure such as balancing workloads. This planning problem is usually called the *machine loading problem*.

The scheduling problems relate to the execution of production orders and include determining the base part input sequencing for each product to be assembled, the product assembly sequence at each station, and the time table for vehicle movements (see Fig. 2.13).

The planning problem and the scheduling problem are associated with different time horizons and are solved sequentially to alleviate the complexity of simultanoeus loading and scheduling. However, in order to best utilize the system capabilities these problems should be solved simultaneously (cf. single-level machine and vehicle scheduling in Sect. 7.3). The integration of these two problems requires some monolithic approach to be applied for solving the corresponding combined resource allocation and scheduling problem. Some recent developments in this direction based on large mixed integer programming formulations are described in Chap. 5.

2.5.1 Machine loading and assembly routing

A flexible assembly system is a mixed-model system (e.g., [39, 64]), i.e., an assembly system in which different models or products are assembled simultaneously. Processing a mix of products makes it possible to use the machines more fully than otherwise. This is because different products spend different amounts of time at the machines, and setup times between task or product changes are negligible.

The main loading objective in a flexible assembly system is to *balance station workloads*, that is, to determine an allocation of assembly tasks and required components among the assembly stations with limited working space such that the total assembly time assigned to each station is equal.

An important difference between machining and assembly systems is that greater capacity of the material handling system is required in a flexible assembly. The assembly times are relatively small and usually shorter than vehicle transfer times. A stop-and-go AGV is sometimes dedicated to each pallet from start of the assembly to finish. Therefore, transportation system in a FAS should have sufficient excess capacity to avoid bottlenecks in the system and its underutilization. In order to eliminate the bottlenecks and to improve the system productivity, the assignment of tasks to assembly stations should be determined based on a multi-objective loading procedure so as to simultaneously balance the workloads of assembly stations and the material handling devices.

In a more general setting the FAS loading is combined with the routing problem to determine for each individual product an *assembly route*, that is, a

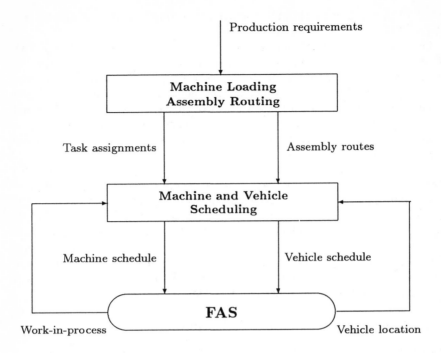

Fig. 2.13. Planning and scheduling in a FAS

sequence of stations the product must visit during its assembly. The objective of the FAS loading and routing is to assign assembly tasks and products to stations so as to equalize the station workloads and to minimize interstation product movements, given the assembly plans selected for a mix of product types to be simultaneously assembled. These two criteria will usually conflict. Minimization of product movements may lead to workloads imbalance, that is, unequal total assembly times assigned to stations and longer intermediate queues and bottlenecks in the system as a result. The less product movements required, the less material handling capacity that must be provided.

The FAS balancing problem is sometimes considered along with the design problem to simultaneously determine the number of stations, the number of parallel assembly machines at each station and the assignment of tasks to stations. Such a joint design and balancing is particularly suitable in the case of a stable mix of product types.

2.5.2 Machine and vehicle scheduling

Production scheduling in a flexible assembly system relates to the detailed sequencing and timing decisions to control the flow of products through the system.

The FAS scheduling decisions are made in a dynamic scheduling environment where various technological and resource dynamic constraints must be considered. An important feature of the FAS scheduling is that both the input and output buffers at stations are limited. Therefore, there is always a possibility that a particular station can be blocked or the system can be locked due to limited buffer spaces. *Machine blocking* occurs when a station cannot move its product to a buffer if the buffer is full, whereas the *system locking* occurs when the system is totally prevented from functioning, i.e., no product movement can be achieved in the system. The limited buffer capacities must be taken into account in scheduling of a FAS.

It is useful to distinguish scheduling decisions that are made off-line from the dispatching decisions that are made in real-time. Real-time scheduling decisions are often based on local considerations as opposed to systemwide off-line scheduling. Similar to the FAS planning decisions, off-line schedule forms a part of pre-production system setup. The results of the off-line scheduling can be next used on-line as a basis for a dispatching decision making.

The scheduling procedures should allow the decisions to be made on-line and in real-time. The complexity of these decisions and the frequency with which they need to be considered implies that most scheduling problems can only be solved heuristically without any guarantee of optimality. However, if alternative assembly modes are available (e.g., through alternative assembly plans or assembly routes) then the difference in system performance under optimal and heuristic solution diminishes (see [120]).

In general, the FAS scheduling can be decomposed into the two basic scheduling subproblems described below.

1. **Base part input sequencing** which refers to sequencing and timing of the release of base parts into the system of various product types to be assembled.

   For each product the time of its relase into the system is determined. The base part of this product is first loaded on a pallet and then the pallet with the base part is loaded on an AGV selected to transport it. The total number of various products to be allowed in the system simultaneously is limited as a result of limited number of available pallets and carriers (AGVs), limited in-process buffers as well as the conditions required to achieve the planned production rate. Too many products in the system may decrease its productivity, owing to congestion and bottlenecks. The selection of a base part to be released into the system and determining its release time is made based on the actual system status (e.g., number of products of each product type in the system, availability of a carrier,

current status and workload of each station, etc.) as well as the decisions made at the planning level (planned station workloads, selected assembly routes, etc.).
2. **Scheduling of assembly and transportation tasks** with the objective of determining an assignment of assembly tasks to station for each individual unit of each product type over a scheduling horizon as well as the associated time table for vehicle movements so as to complete a production order and to minimize some optimality criterion. The flow of products through the system is a result of implementation the assembly and transportation tasks scheduling decisions.

If on-line scheduling of assembly and transportation tasks is applied, then various dispatching decisions are made in real-time. For example, such decisions may include:
— selection of next task for assignment to an idle station from the set of products waiting in its input buffer;
— selection of product for transfer to another station from the set of products waiting in a station output buffer;
— selection of the destination station to perform next task for the product selected for transfer;
— selection of an AGV to transfer product between stations.

The above described two basic scheduling problems are strictly interrelated. The base part input sequence depends on the current state of assembly process, which in turn is a result of the assembly and transportation task schedule and the actual product flow control. On the other hand, the assembly and transportation task schedule is constrained by the base part input sequence which controls the release of new parts into the system.

2.5.3 Planning and scheduling in electronics assembly

Production planning and scheduling is particularly complicated in the electronics industry, owing to a large variety of product types that are assembled of a great number of different components using alternative assembly plans and assembly routes. For example, in the assembly of printed wiring boards at the Texas Instruments facility in Austin, Texas, planning and scheduling is complicated by the need to deal with over 10,000 different components and up to 400 different board types, each with their individual routing and bill of materials. In order to improve the real-time scheduling of generic assembly acitivities an intelligent decision support system INSITES – *Integrated Scheduling, Inventory and Throughput Evaluation System* ([35]) has been implemented. The system uses a family of intelligent heuristics to quickly generate a set of schedules that can be adjusted to meet the current objectives of the company with regard to on-time performance, net revenue and workload balance. The following scheduling objectives were considered at TI ([35]):

2.5 Planning and scheduling

- meeting due dates for boards;
- balancing workloads over facility resources;
- reducing waiting times;
- maximizing overall throughput and productivity;
- simplifying and standardization the task of daily scheduling;
- allowing each station manager and daily scheduling manager the greatest amount of autonomy.

In printed circuit board assembly where the most sophisticated placement machines such as Fuji CP II are used (see, Sect. 1.3.2), the FAS planning and scheduling is usually combined with a detailed optimization of an assembly process on each computer numerically controlled machine. For example, in order to improve system productivity and to increase the production rate in the PCB assembly using the Dynapert MPS 500 placement machine (see Fig. 1.10 in Sect. 1.3.2), a hierarchical framework consisting of the following optimization problems is identified in [3, 5].

1. The component allocation problem: to determine the number of feeders of each component type assigned to a machine.
2. The partitioning problem: to assign each component to one of the two carriers along with the corresponding sets of nozzles.
3. The placement sequencing problem: to specify the sequence of component placements, taking into consideration 2.
4. Reel positioning problem: to assign reels of components to slots on the carriers, given the assignment of components to carriers and the placement sequence.

A similar, hierarchical approach is proposed in [20] to optimize the PCB assembly process using the Fuji CP II placement machine, see Fig. 1.12. The assembly process can be decomposed into the following three subproblems.

1. Table tour – the placement sequence is specified to determine the movement of the table.
2. Component assignment – the assignment of component types to magazine slots is determined.
3. Component retrieval sequence – the movement of the magazine rack is determined if more than one slot is assgned the same component type.

For the given placement sequence, the component assignment problem and the component retrieval sequence problem are collectively known as the *component retrieval problem* in PCB assembly, see Sect. 5.2. For more details on the problem formulations and solution algorithms, see [20, 26, 54].

The various planning problems and solution approaches are discussed in Chap. 3 and 4, whereas the scheduling problems and algorithms for various FAS configurations are presented in Chap. 5, 6 and 7.

3. Loading and Routing Decisions in Flexible Assembly Systems

This chapter is devoted to loading and routing in flexible assembly systems. The loading and routing are considered as system set up decision making problems that must be solved before a FAS can begin to assemble products. The solution specifies that all component feeders and assembly tools required for each assembly task are located at the appropriate stations and that each product is successively routed to the selected stations subject to precedence relations among the assembly tasks. (An assembly route is defined to be a sequence of stations that a product must successively visit to have all its required components assembled with the base part.)

The loading and routing decisions immediately precede detailed scheduling of an assembly process and they are not changed during the entire processing horizon until all product requirements are completed. Once the loading and routing problem has been solved, an assembly schedule will be obtained by solving the corresponding machine and vehicle scheduling problem in a convenient way, see Chap. 5, 6 and 7.

The objective of the FAS loading and routing is to assign assembly tasks and products to stations so as to equalize the station workloads and to minimize interstation product movements subject to precedence relations among the tasks for a mix of product types. These two criteria will usually conflict. Minimization of product movements may lead to workload imbalance, that is, unequal total assembly times assigned to stations and longer intermediate queues and bottlenecks in the system as a result. The less product movements required, the less material handling capacity that must be provided. In order to best utilize the system capabilities the FAS loading problem should simultaneously account for both the assembly times and the transportation times between the stations. The assembly times are relatively small and usually shorter than vehicle transfer times. As a result a stop-and-go AGV is often dedicated to each pallet from start of the assembly to finish. Therefore, transportation systems should have sufficient excess capacity to avoid bottlenecks in the system and its underutilization. Furthermore, when transportation times are greater than assembly times, neglecting the former at the machine loading level may lead to bottlenecks on some paths of the transportation network. Transportation times can also contribute to station idle time if stations have to wait for the delivery of the next product for assembly.

The workload balance can be defined in several ways. In [121] it is shown that there is a unique *continuous* workload distribution that maximizes throughput of a system. On the other hand there is no universally accepted measure of workload balance for loading problems where the task assignments are *discrete*. Researchers use different surrogate objective functions for the same stated objective of balancing workloads. For example, in [7] the objective of minimizing the maximum difference between the average and the actual station workload is used. In [53] minimization of the difference between maximum and minimum station workloads is proposed. In [61] 12 surrogate objective functions as measures of workload balance were investigated. In [1] and [16] a workload distribution is defined to be balanced if it minimizes the workload of the busiest station. The latter definition is mainly used throughout this chapter.

Loading and routing problems are considered by many researchers. For example, in [1] and [9] workload balancing and part transfer minimization is formulated in terms of a particular capacity assignment problem with continuous variables and with no limitations on working space available. In [2] a linear program and an efficient column generation technique is proposed to find a routing mix that minimizes total transportation time, given a constraint on each station workload and no constraint on station working space.

The FAS balancing problem is often considered along with the design problem to simultaneously determine the number of stations, the number of parallel assembly machines at each station and the assignment of tasks to stations. Such joint design and balancing is particularly suitable where the product mix is expected to be stable, e.g., [63].

In this chapter various single and bi-objective integer programming formulations are proposed for the loading and routing in two types of flexible assembly systems: flexible assembly lines (FAL) and general flexible assembly systems. A FAL is a unidirectional flow system where a base part enters the system and is processed by a series of stations. A part may bypass some stations but does not revisit any station. In a general FAS, however, direction of flow varies for different products and revisiting of stations is allowed. While integer programming formulations have been widely used to express the assembly line balancing problem (see [39]), their application in FAS design and balancing is not so extensive. Examples of the integer programming formulations for the latter problem can be found in [1, 42, 61]. In [42], 0-1 integer programs are presented to assign tasks to workstations and to select assembly equipment for each workstation. These integer programming models do not take into account the aspects of product flow or of the assembly machine parallelism at each workstation. Duplicate assignment of tasks to stations and alternative routings available for each product type with both the assembly and transportation times accounted for are the other factors which should be introduced into the integer programming models.

The integer programming models proposed in this chapter can be classified into the following three categories (e.g., [102]):

A. Models with fixed assembly routes, where each assembly task is assigned to only one station
   1. Models of flexible assembly lines: unidirectional flow without revisiting of stations
   2. Models of general flexible assembly systems: multidirectional flow with revisiting of stations allowed
B. Models with alternative assembly routes, where each assembly task can be assigned to more than one machine
   1. Models of flexible assembly lines with parallel machines: unidirectional flow without revisiting of stations
   2. Models of general flexible assembly systems: multidirectional flow with revisiting of stations allowed
C. Models of FAS design and balancing for a stable mix of product types.

Models of category **A** and **B** are presented in Sect. 3.2, 3.5 and 3.6, and models of category **C** in Sect. 3.3. In Sect. 3.2 and 3.3 a family of integer programming models for the FAS design and balancing are presented for a variety of different system configurations and objective functions. In Sect. 3.5 and 3.6 integer programming formulations are presented for the biobjective optimization of station workloads and station-to-station product movements, and in Sect. 3.5.2 and 3.6.2 two different solution approaches are proposed. In Sect. 3.5.2 an interactive heuristic for simultaneous loading and routing is presented, and in Sect. 3.6.2 a lexicographic approach with a linear relaxation-based heuristic is described for sequential loading and routing. Numerical examples illustrating various possible applications of the models and the solution approaches proposed are provided in Sect. 3.4, 3.5.3 and 3.6.4.

3.1 Description of a flexible assembly system

A flexible assembly system (FAS) consists of a set of assembly stations and a loading/unloading (L/U) station connected by conveyors or transporter paths. In the system several different product types can be assembled simultaneously. A typical assembly process proceeds as follows. A base part of an assembly is loaded on a pallet and enters the FAS at the L/U station. As the pallet is carried by conveyors or automated guided vehicles (AGVs) through assembly stations, components are assembled with the base part. When all the required components are assembled with the base part, it is carried back to the L/U station and the complete assembly leaves the FAS.

Each product is assembled according to an assembly plan selected for this product. An assembly plan (see Sect. 2.4) is defined to be a set of precedence relations among the assembly tasks required to complete the product. A

44   3. Loading and Routing Decisions in Flexible Assembly Systems

typical assembly plan can be represented in the form of an assembly tree or an assembly chain, where the latter is often called an assembly sequence [21, 55].

The flexible assembly machines (e.g., assembly robots or automatic insertion machines) have a finite working space due to their physical configuration. The component feeding mechanism associated with each assembly task uses some of the finite working space. Therefore only a limited number of tasks can be assigned to a robot (see Fig. 1.4).

When components are all of relatively similar sizes one may assume that each task uses the same amount of the station working space. Under this assumption the finite working space of a robot can be refined as its flexibility capacity [64] which specifies the maximum number of tasks that can be assigned to the robot. There are negligible setup times between task changes among the tasks assigned to a robot, e.g., [63, 64].

Let us consider a FAS made up of m assembly stations $i \in I = \{1, \ldots, m\}$ and the L/U station connected by AGV paths. The configuration of the station layout is a general network. In the system n different types of assembly tasks $j \in J = \{1, \ldots, n\}$ can be performed to simultaneously assemble v products $k \in K = \{1, \ldots, v\}$ of various types. Let $I_j \subset I$ be the subset of stations capable of performing task j. Each station $i \in I$ has a finite working space b_i where a limited number of component feeders and gripper magazines can be placed. As a result only a limited number of assembly tasks can be assigned to one assembly station. Let a_{ij} be the amount of station $i \in I_j$ working space required for task j.

Each product $k \in K$ requires a subset J_k of assembly tasks to be performed subject to precedence relations. The precedence relations for product k are defined by the set R_k of immediate predecessor-successor pairs of assembly tasks (j, r) such that task $j \in J_k$ must be performed immediately before task $r \in J_k$.

Finally, denote by p_{jk} the assembly time required for task $j \in J_k$ of product k and by q_{il} the transportation time required to transfer a product from station i to station l.

3.2 Optimization of station workloads and product movements

In order to best utilize the system capabilities and to minimize the completion time of all product requirements, the FAS balancing problem should simultaneously account for both the assembly times and the transportation times between the stations.

The models presented in this section aim at simultaneous optimization of station workloads and station-to-station product flows. The problem objective is to determine the optimal assignment of tasks to station and the assembly routes for all products so as to minimize the sum of weighted assembly and transportation time.

The solution of each model must satisfy the following four basic types of constraints:

- The demand for all products must be satisfied.
- The number of tasks assigned to each station must not exceed its flexibility capacity.
- Tasks must be assigned to stations such that precedence relations among the tasks are maintained with no revisiting of stations required in a FAL.
- Products must be routed to the stations where the required assembly tasks have been assigned.

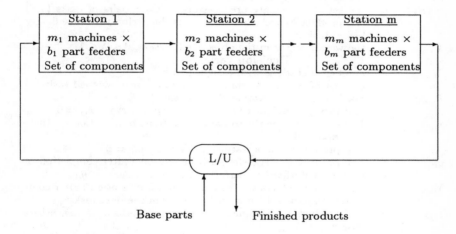

Fig. 3.1. A flexible assembly line with parallel machines

Models **M1** and **M2** are proposed for flexible assembly lines, in which products do not revisit any station. In the model **M1** each station is a single assembly machine, whereas in **M2** a station consists of identical parallel machines (see Fig. 3.1). Model **M3** is constructed for a general FAS, in which revisiting of stations is allowed.

In the models **M1** and **M2** precedence relations among the assembly tasks for all the product types are merged and are represented by a single super-precedence constraint, whereas in the model **M3** individual precedence constraints for each product type are introduced. The notation used in the models proposed is shown in Table 3.1.

In order to simplify the presentation of models proposed for flexible assembly lines, all machines in the FAL are assumed to be of the same type, which is flexible enough to perform all assembly tasks. This assumption can

Table 3.1. Notation

		Indices
i	=	assembly station, $i \in I = \{1, \ldots, m\}$
j	=	assembly task, $j \in J = \{1, \ldots, n\}$
h	=	parallel machine at station i, $(h = 1, \ldots, m_i)$
k	=	product, $k \in K = \{1, \ldots, v\}$
		Input parameters
a_{ij}	=	working space required for assignment of task j to station i
b_i	=	total working space of station i (number of tasks that may be assigned to station i, if all $a_{ij} = 1$)
d_k	=	demand for product k
m_i	=	number of parallel machines at station i
p_{jk}	=	assembly time for task j of product k
q_{il}	=	transportation time from station i to station l
I_j	=	the set of stations capable of performing task j
J_k	=	the set of tasks required for product k
$R\,(R_k)$	=	the set of immediate predecessor-successor pairs of tasks (j, r) (for product k) such that task j must be performed immediately before task r
λ	=	the weight factor in the objective function, $0 \leq \lambda \leq 1$
		Decision variables
M	=	number of stations with assigned tasks
M_i	=	number of parallel machines at station i with assigned tasks
P_{max}	=	maximum station workload (cycle time)
x_{ij}	=	1, if task j is assigned to station $i \in I_j$; otherwise $x_{ij} = 0$
x_{ihj}	=	1, if task j is assigned to parallel machine h at station i; otherwise $x_{ihj} = 0$
y_{ik}	=	1, if product k visits station i; otherwise $y_{ik} = 0$
y_{iljk}	=	1, if product k is transferred from station i after completion of task j, to station l to perform next task; otherwise $y_{iljk} = 0$
Y_{iljk}	=	number of products k to be transferred from station i after completion of task j, to station l to perform the next task
z_{ijk}	=	1, if product k is assigned to station i to perform task j; otherwise $z_{ijk} = 0$
Z_{ijk}	=	number of products k assigned to station i to perform task j

be easily relaxed to extend the models for multiple machine types. When a machine i is incapable of performing some tasks j, (i.e., $i \notin I_j$), the corresponding assignment and routing variables are restricted to zero.

Another simplified assumption made in this section is that of equal working space required for all tasks. Under this assumption, the finite working space b_i of a station can be redefined as its flexibility capacity, i.e., the maximum number of different tasks that can be assigned to a station. If the latter assumption does not hold, then different working space requirement coefficients a_{ij} for each task should be introduced into the corresponding constraints with the b_i's coefficients defined again as a station working space available.

Model M1: *Balancing station workloads and minimizing product movements in a flexible assembly line*
Minimize
$$F_{max} = \lambda P_{max} + (1-\lambda) \sum_{i \in I} \sum_{k \in K} d_k y_{ik} \qquad (3.1)$$

subject to

$$\sum_{k \in K} \sum_{j \in J_k} d_k p_{jk} x_{ij} \leq P_{max}; \qquad i \in I \qquad (3.2)$$

$$\sum_{i \in I} x_{ij} = 1; \qquad j \in J \qquad (3.3)$$

$$\sum_{j \in J} x_{ij} \leq b_i; \qquad i \in I \qquad (3.4)$$

$$\sum_{i \in I} i x_{ij} \leq \sum_{i \in I} i x_{ir}; \qquad (j,r) \in R \qquad (3.5)$$

$$y_{ik} \geq x_{ij}; \qquad i \in I, k \in K, j \in J_k \qquad (3.6)$$

$$x_{ij} \in \{0,1\}; \qquad \forall i,j \qquad (3.7)$$

$$y_{ik} \in \{0,1\}; \qquad \forall i,k \qquad (3.8)$$

The first term of the objective function F_{max} (3.1) corresponds to station workload balance and the second term to station visits by all products. The weight factor λ ($0 \leq \lambda \leq 1$) is used for trading-off between the two criteria. Constraint (3.2) determines workload P_{max} of the bottleneck station. Constraint (3.3) ensures that each task is assigned to only one station. The station flexibility capacity constraint (3.4) guarantees that the total number of tasks assigned to a station does not exceed the number of available part feeders. Constraint (3.5) maintains the precedence relations among the assembly tasks and ensures that a product does not revisit any station in a unidirectional flow system. If task r is assigned to station i, then task of each type j, that is to be performed immediately before r must be assigned to a station l such that $l \leq i$. Constraint (3.6) ensures that each product is routed to the station where the required tasks are assigned.

Model M2: *Balancing station workloads and product movements in a flexible assembly line with parallel machines*
Minimize
$$F_{max} \qquad (3.9)$$

subject to

$$\lambda \sum_{k \in K} \sum_{j \in J_k} (d_k p_{jk} x_{ij}/m_i) + (1-\lambda) \sum_{k \in K} d_k y_{ik} \leq F_{max}; \quad i \in I \qquad (3.10)$$

$$\sum_{i \in I} x_{ij} = 1; \quad j \in J \qquad (3.11)$$

$$\sum_{j \in J} x_{ij} \leq m_i b_i; \quad i \in I \qquad (3.12)$$

$$x_{ij} \leq \sum_{j \in J}(x_{i-1j}/m_{i-1}); \quad \forall i > 1, j \in J \qquad (3.13)$$

$$\sum_{i \in I} i x_{ij} \leq \sum_{i \in I} i x_{ir}; \quad (j, r) \in R \qquad (3.14)$$

$$y_{ik} \geq x_{ij}; \quad i \in I, k \in K, j \in J_k \qquad (3.15)$$

$$x_{ij} \in \{0, 1\}; \quad \forall i, j \qquad (3.16)$$

$$y_{ik} \in \{0, 1\}; \quad \forall i, k \qquad (3.17)$$

The objective function F_{max} (3.9) represents the weighted sum of assembly times and product movements to the bottleneck station defined by constraint (3.10). The first term of constraint (3.10) models workload of the bottleneck station, and the second term is total number of the station visits by products that require the tasks assigned to this station. The second term can be considered to be the total transfer time to the bottleneck station in the case of equal transportation times.

Constraint (3.12) accounts for the total flexibility capacity of all parallel machines at each station. Constraint (3.13) ensures that tasks are not assigned to station i unless at least one task has been assigned to each machine of station $i-1$. Constraints (3.11), (3.14) and (3.15) have similar functions as in model **M1**.

In the following model **M3** of a general flexible assembly system revisiting of stations is allowed and assembly process of each product is subject to individual precedence constraints. Model **M3** also accounts for the product transfer times between the stations.

Model M3: *Balancing total assembly and transportation time in a general FAS*

Minimize

$$F_{max} \qquad (3.18)$$

subject to

$$\lambda \sum_{k \in K} \sum_{j \in J_k} d_k p_{jk} x_{ij} + (1-\lambda) \sum_{k \in K} \sum_{l \neq i} \sum_{j \in J_k} d_k q_{li} y_{lijk} \leq F_{max}; \quad i \in I \qquad (3.19)$$

$$\sum_{i \in I_j} x_{ij} = 1; \quad j \in J \qquad (3.20)$$

$$\sum_{j \in J} x_{ij} \leq b_i; \quad i \in I \qquad (3.21)$$

$$x_{ij} + x_{lr} - y_{iljk} \leq 1; \quad k \in K, i \in I_j, l \in I_r, l \neq i, (j, r) \in R_k \qquad (3.22)$$

$$-x_{ij} - x_{lr} + 2y_{iljk} \leq 0; \quad k \in K, i \in I_j, l \in I_r, l \neq i, (j, r) \in R_k \qquad (3.23)$$

$$x_{ij} \in \{0, 1\}; \quad \forall i, j \qquad (3.24)$$

$$y_{iljk} \in \{0, 1\}; \quad \forall i, l, k \qquad (3.25)$$

3.2 Optimization of station workloads and product movements

The objective function F_{max} (3.18) represents the weighted sum of total assembly and transportation time for the bottleneck station defined by constraint (3.19). The first term of constraint (3.19) denotes the workload of the bottleneck station, and the second term is the total transportation time required to transfer products to this station.

Constraint (3.20) ensures that each task is assigned to only one station, and (3.21) is the flexibility capacity constraint.

Constraints (3.22) and (3.23) ensure that each product is routed to the stations where the required tasks are assigned. Implicitly, (3.22) and (3.23) account for the individual precedence constraints of each product type.

Models **M1** and **M2** contain $m(n+v)$ binary variables x_{ij}, y_{ik} and generate $mnv + 2m + n^2$ constraints, whereas in the model **M3** there are $mn + m(m-1)(n-1)v$ binary variables x_{ij}, y_{iljk} and $2m(m-1)n(n-1) + 2m + n$ constraints.

It is evident that model **M3** leads to a larger sized integer program than models **M1** and **M2**. However, it allows the individual precedence constraints for each product type to be considered and detailed assembly routes to be determined with transfer times between the stations taken into account.

Note that due to a special network flow structure of constraint (3.6), variables y_{ik} in model **M1** need not to be binary constrained. If y_{ik} are allowed to be continuous, the solution generated by this relaxation always has binary values of y_{ik}.

In the models presented so far only **M2** allows for alternative routes due to duplicate assignment of a task to several parallel machines at one station. However, none of the models **M1**, **M2**, **M3** allowed the duplicate assignment of a task to more than one station so that alternative assembly routes could be available in the system without parallel machines.

Duplicate assignment of tasks to stations and the corresponding alternative routes for each product type with possible revisiting of stations are allowed in the model **M4** presented below.

Model M4: *Balancing station workloads in a FAS with alternative assembly routes*

Minimize

$$P_{max} \qquad (3.26)$$

subject to

$$\sum_{k \in K} \sum_{j \in J_k} p_{jk} Z_{ijk} \leq P_{max}; \qquad i \in I \qquad (3.27)$$

$$\sum_{i \in I_j} x_{ij} \geq 1; \qquad j \in J \qquad (3.28)$$

$$\sum_{j \in J} x_{ij} \leq b_i; \qquad i \in I \qquad (3.29)$$

$$x_{ij} \leq \sum_{l=i}^{m} x_{lr}; \quad i \in I_j,\ (j,r) \in R \qquad (3.30)$$

$$x_{lr} \leq \sum_{i=1}^{l} x_{ij}; \quad l \in I_r,\ (j,r) \in R \qquad (3.31)$$

$$\sum_{i \in I_j} Z_{ijk} = d_k; \quad k \in K,\ j \in J_k \qquad (3.32)$$

$$x_{ij} \leq \sum_{\{k:j \in J_k\}} Z_{ijk} \leq \Big(\sum_{\{k:j \in J_k\}} d_k\Big) x_{ij}; \quad i \in I_j,\ j \in J \qquad (3.33)$$

$$x_{ij} \in \{0,1\}; \quad \forall i,j \qquad (3.34)$$

$$Z_{ijk} \geq 0,\ \text{integer}; \quad \forall i,j,k \qquad (3.35)$$

The objective function (3.26) is a measure of system imbalance, and constraint (3.27) defines workload P_{max} of the bottleneck station. Constraint (3.28) ensures that each task is assigned to at least one station. and (3.29) is the flexibility capacity constraint. Constraints (3.30) and (3.31) ensure reaching of such a duplicate assignment of tasks to stations that revisiting of stations by products could be avoided. However, in order to balance station workloads one has to allow assembly routes with varying direction of product flow. Constraint (3.32) guarantees that the demand for all products is satisfied, and (3.33) ensures that each product is routed to the stations where the required tasks are assigned.
Model **M4** requires $(v+1)mn$ decision variables and generates $(v+1)n + 2m(n^2+1)$ constraints.

3.3 Design and balancing of flexible assembly lines

In this section two models are proposed for simultaneous balancing of station workloads and minimization of the number of stations and/or the number of all assembly machines in a flexible assembly line. Model **M5** is proposed for the flexible assembly line with a single machine at each station, whereas model **M6** is dedicated to the line with parallel machines.

Model M5: *Design and balancing a flexible assembly line*
Minimize
$$\lambda P_{max} + (1-\lambda) M \qquad (3.36)$$

subject to

$$\sum_{k \in K} \sum_{j \in J_k} p_{jk} x_{ij} \leq P_{max}; \quad i \in I \qquad (3.37)$$

$$\sum_{i \in I} x_{ij} = 1; \quad j \in J \qquad (3.38)$$

$$\sum_{j \in J} x_{ij} \leq b_i; \qquad i \in I \qquad (3.39)$$

$$\sum_{i \in I} i x_{ij} \leq \sum_{i \in I} i x_{ir}; \qquad (j, r) \in R \qquad (3.40)$$

$$i x_{ij} \leq M; \qquad i \in I, j \in J \qquad (3.41)$$

$$x_{ij} \in \{0, 1\}; \qquad \forall i, j \qquad (3.42)$$

The objective function (3.36) models the trade-off between the minimum cycle time P_{max} and the minimum number of assembly stations M. If $\lambda = 0$, then (3.36) minimizes the number of all stations, and if $\lambda = 1$, (3.36) minimizes the cycle time.

Given the cycle time P_{max}, the minimum number of stations M is constrained as follows

$$\sum_{k \in K} \sum_{j \in J_k} p_{jk}/P_{max} \leq M \leq n,$$

with

$$n \leq \sum_{i=1}^{M} b_i < n + b_M.$$

Constraint (3.37) guarantees that each station workload does not exceed the cycle time P_{max}, and (3.41) defines the number M of the last station with assigned tasks. Constraints (3.38), (3.39) and (3.40) have similar functions as in the previous models.

Model M6: *Design and balancing a flexible assembly line with parallel machines*

Minimize

$$\lambda P_{max} + (1 - \lambda) \sum_{i \in I} M_i \qquad (3.43)$$

subject to

$$\sum_{k \in K} \sum_{j \in J_k} p_{jk} x_{ihj} \leq P_{max}; \qquad i \in I, h = 1, \ldots, m_i \qquad (3.44)$$

$$\sum_{i \in I} x_{ij} = 1; \qquad j \in J \qquad (3.45)$$

$$\sum_{j \in J} x_{ij} \leq m_i b_i; \qquad i \in I \qquad (3.46)$$

$$x_{ij} \leq \sum_{h=1}^{m_i} x_{ihj} \leq m_i x_{ij}; \qquad i \in I, j \in J \qquad (3.47)$$

$$\sum_{j \in J} x_{ihj} \leq b_i; \qquad i \in I, h = 1, \ldots, m_i \qquad (3.48)$$

$$\sum_{i \in I} i x_{ij} \leq \sum_{i \in I} i x_{ir}; \qquad (j, r) \in R \qquad (3.49)$$

$$hx_{ihj} \leq M_i; \quad i \in I,\ h=1,\ldots,m_i,\ j \in J \quad (3.50)$$
$$M_i/m_i \leq M_{i-1}/m_{i-1}; \quad \forall i > 1 \quad (3.51)$$
$$x_{ij} \in \{0,1\}; \quad \forall i,j \quad (3.52)$$
$$x_{ihj} \in \{0,1\}; \quad \forall i,h,j \quad (3.53)$$

Similarly as in **M5** the objective function (3.43) models the trade-off between the minimum cycle time P_{max} and the minimum number of machines $\sum_{i \in I} M_i$ with assigned tasks.
Constraint (3.44) guarantees that each station workload does not exceed the cycle time P_{max}. Constraint (3.47) ensures that each task is assigned to at least one machine and not more than all m_i parallel machines of such a station i where the task has been assigned. Constraint (3.48) guarantees that the number of tasks assigned to each machine does not exceed the number of part feeders available. Constraint (3.50) defines for each station i the number M_i of the last parallel machine with assigned tasks. Constraint (3.51) does not allow to assign tasks to next station unless all parallel machines at immediately preceding station have been assigned tasks. Constraints (3.45), (3.46) and (3.49) have similar functions as in model **M2**.

If the cycle time P_{max} is minimized only ($\lambda = 1$), then constraints (3.50) and (3.51) with variables M_i should be deleted.

Model **M5** contains mn binary variables x_{ij}, one integer variable M and generates $2m + mn + n^2$ constraints, whereas in the model **M6** there are $(m + \sum_{i=1}^{m} m_i)n$ binary variables x_{ij}, x_{ihj}, m integer variables M_i and $2m + (n+2)\sum_{i=1}^{m} m_i + mn + n^2$ constraints.

3.4 Numerical examples

In this section several numerical examples are presented to illustrate various applications of the models proposed. The models have been applied for the design and balancing of a hypothetical FAS, in which $n = 15$ component types are assembled to produce $v = 4$ product types. The corresponding ordered sequences of tasks $j \in J_k$ required to make each product type $k = 1,2,3,4$ are the following: (1,2,3,4,6,12,14,15), (1,2,5,6,9,10,13,15), (2,4,5,7,8,9,10,14), (8,11,13,14,15).
The resulting super-precedence constraints for all the products are represented by the following set of consecutive pairs of tasks:
$R = \{(1,2),(2,3),(2,4),(2,5),(3,4),(4,5),(4,6),(5,6),(5,7),(6,9),(6,12),(7,8),(8,9),(8,11),(9,10),(10,13),(10,14),(11,13),(12,14),(13,14),(13,15),(14,15)\}$.
Graph of super-precedence relations among the tasks is shown in Fig. 3.2.
Assembly times p_{jk} for each task j are identical for all product types, (i.e., $p_{jk} = p_j, \forall k,\ j \in J_k$) and are as follows: $p_1 = 4$, $p_2 = 2$, $p_3 = 2$, $p_4 =$

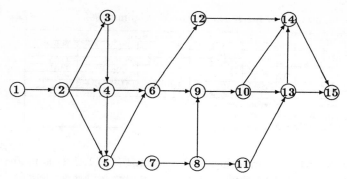

Fig. 3.2. Graph of super-precedence relations

2, $p_5 = 4$, $p_6 = 2$, $p_7 = 3$, $p_8 = 5$, $p_9 = 2$, $p_{10} = 4$, $p_{11} = 5$, $p_{12} = 2$, $p_{13} = 4$, $p_{14} = 2$, $p_{15} = 3$.

First, the models **M1, M2, M3** with various values of the weight factor λ and model **M4** have been applied to determine station workloads and assembly routes in a FAS which is made up of $m = 3$ stations. In **M1, M2, M3** each station i ($i = 1, 2, 3$) has $b_i = 5$ part feeders, and in **M4**, $b_i = 7$ part feeders. Each product type k ($k = 1, 2, 3, 4$) has the same demand of $d_k = 100$ units. The system is organized as a series of the stations 1,2,3 along an AGV path. The transfer time between the successive stations is equal to 2 time units, and hence transportation times q_{il} from station i to station l are: $q_{12} = q_{21} = 2$, $q_{13} = q_{31} = 4$, $q_{23} = q_{32} = 2$.

Table 3.2. Task assignments and assembly routes

Model (var,con)	Task assignments			Assembly routes for products type			
	Station 1	Station 2	Station 3	1	2	3	4
M1 (57,132)	(a) 1-5	6-10	11-15	1,2,3	1,2,3	1,2,3	2,3
	(b) 1-5	6-10	11-15	1,2,3	1,2,3	1,2,3	2,3
	(c) 1-5	6-10	11-15	1,2,3	1,2,3	1,2,3	2,3
M2 (57,161)	(a) 1-5,7	6,12	8-11,13-15	1,2,3	1,2,3	1,3	2,4
	(b) 1-5,7	6,8-10,12	11,13-15	1,2,3	1,2,3	1,2,3	2,3
	(c) 1-5	6-10	11-15	1,2,3	1,2,3	1,2,3	2,3
M3 (195,322)	(a) 1-5	6-10	11-15	1,2,3	1,2,3	1,2,3	2,3
	(b) 11-15	6-10	1-5	3,2,1	3,2,1	3,2,1	2,1
	(c) 3,8,9,11,13	2,5,6,7,10	1,4,12,14,15	3,2,1,3,2,3	3,2,1,2,3	2,3,2,1,2,3	1,3
M4 (133,273)	1-6	5-10,12	10,11,13-15	1,2,3	1,2,3 (91)(d) 1,2,3 (9)(d)	1,2 (84)(d) 1,2 (16)(d)	2,3

(a) $\lambda = 0$
(b) $\lambda = 0.5$
(c) $\lambda = 1$
(d) number of products assigned to the route: $Z_{2,10,2} = 91$, $Z_{3,10,2} = 9$, $Z_{2,10,3} = 84$, $Z_{3,10,3} = 16$
(var,con) – number of variables and number of constraints

The following three variants of the balancing problem have been considered in which the objective function accounts for: total transportation time only ($\lambda = 0$), total weighted assembly and transportation time ($\lambda = 0.5$), and total assembly time only ($\lambda = 1$).

3. Loading and Routing Decisions in Flexible Assembly Systems

Table 3.3. Cycle time, number of machines and task assignments

Model (var,con)	λ	P_{max}, $M^{(a)}$	Task assignments				
			Station 1	Station 2	Station 3	Station 4	Station 5
M5 (76,123)	0	30, 3	1-5	6-10	11-15	-	-
	0.5	20, 5	1,2	3,4,5,6	7,8,11,12	9,10,13	14,15
	1	20, 5	1,2,3,4	5,6,7,12	8,11	9,10,13	14,15
M6 (231,367)	0	30, 3	1-10	11-15	-	-	-
	0.5	14, 8	1,2,3,4	5,7,8	6,9,10,12	11,13,14,15	-
(226,213)	1(b)	12, 9	1	2,3,4,5,6	7,8,11,12	9,10,13	14,15

(a) $M = \sum_i M_i$ – total number of machines with assigned tasks
(b) model M6 without constraints (3.50) and (3.51)
(var,con) – number of variables and number of constraints

In addition, the model **M2** assumes that each station i ($i = 1, 2, 3$) is made up of $m_i = 2$ identical parallel machines, each having $b_i = 5$ part feeders. The solution results are shown in Table 3.2. The solutions obtained for model **M1** are identical for all λ, independently of the weighting of the two objectives. This is a result of predominated constraints (3.4) and (3.5) for the example data. Such a behaviour is not observed for the other models.

Model **M4** selects two similar routes for each product of type 2 and 3, however the routes are not identical. Component $j = 10$ is assembled at two different stations 2 and 3 for a specified number of products, (see variables Z_{ijk} in Table 3.2).

Comparisons among the results obtained for the various models proposed can be difficult for many reasons. Models **M1** and **M2** account for one set of the super-precedence relations for all product types and do not allow revisiting of stations. Model **M3** however, accounts for the individual precedence relations for each product type and detailed transfer times between the stations with revisiting allowed. Finally, model **M4** admits duplicate assignment of tasks to stations and determines alternative assembly routes with possible revisiting of stations.

Next, the models **M5** and **M6** have been applied for the simultaneous design and balancing of a flexible assembly line which can be made up of at most $m = 5$ stations each supplied with $b_i = 5$ ($i = 1, \ldots, m$) part feeders. In addition, in the model **M6** each station i can be organized as $m_i = 2$ identical parallel machines, each having 5 part feeders.

The solution results are shown in Table 3.3. The solutions obtained for model **M5** with $\lambda = 0.5$ and $\lambda = 1$ yield the same cycle time and the number of selected machines with different task assignments. This is a result of alternate optima, which exist for the example data.

Note that models **M1**, **M5** and **M6** can be considered to be reduced formulations of bicriterion problems, where the weight λ represents the relative importance of original criteria. Hence, the corresponding results shown in the tables are nondominated solutions (weakly nondominated for $\lambda = 0, 1$) of the original bicriterion problems.

The examples presented in this section were solved using discrete optimizer LINGO [115], in which a branch and bound method is applied. CPU run times on a PC 486 have been varied from several seconds to over one

3.4 Numerical examples

```
MODEL M3:
1] SETS:
2] !ASSEMBLY STATIONS WITH LIMITED STAGING CAPACITY S(I);
3] MA / @FILE(M3.LDT)/: S;
4] !OPERATIONS (PART TYPES);
5] OP / @FILE(M3.LDT)/;
6] !ASSEMBLY TYPES WITH DEMANDS D(K);
7] AS / @FILE(M3.LDT)/: D;
8] !SET JK(K,J) OF REQUIRED OPERATIONS J FOR ASSEMBLY TYPE K;
9] JK( AS, OP) / @FILE(M3.LDT)/;
10] !PRECEDENCE RELATIONS PRED(K,J,R) FOR ASSEMBLY TYPE K;
11] PRED(AS, OP, OP) / @FILE(M3.LDT)/ ;
12] !SET OF FINAL OPERATIONS UL(K) FOR ASSEMBLY TYPE K;
13] UL(AS, OP) / @FILE(M3.LDT)/ ;
14] !PROCESSING TIMES P(J,K);
15] PT( OP, AS): P;
16] !TRANSPORTATION TIMES Q(I,L);
17] TT( MA, MA): Q;
18] !DECISION VARIABLES: OPERATION ASSIGNMENTS;
19] ASS( MA, OP): X;
20] !DECISION VARIABLES: INTERMACHINE MOVEMENTS OF ASSEMBLIES;
21] FLOW( MA, MA, OP, AS): Y;
22] ENDSETS
23]
24] !THE MODEL;
25]
26] MIN = FMAX;
27]
28] !SUBJECT TO;
29]
30] !EACH OPERATION (PART TYPE) MUST BE ASSIGNED TO ONLY ONE STATION;
31] @FOR( OP(J): @SUM( MA(I): X( I, J)) = 1);
32]
33] !STAGING CAPACITY CONSTRAINTS;
34] @FOR( MA(I): @SUM( OP(J): X( I, J)) < S(I));
35]
36] !MAXIMUM WEIGHTED SUM FMAX OF PROCESSING AND TRANSPORTATION TIME;
37] @FOR( MA(I): LAMBDA * @SUM( AS(K): @SUM( OP(J)| @IN( JK, K, J):
38] D(K)* P( J, K)* X( I, J))) + (1 - LAMBDA) * @SUM( AS(K): @SUM( MA(L)| L #NE# I:
39] @SUM( OP(J)| @IN( JK, K, J): D(K)* Q( L, I)* Y( L, I, J, K)))) < FMAX);
40]
41] !ROUTE SELECTION CONSTRAINTS;
42] @FOR( AS(K): @FOR( MA(I): @FOR( MA(L)| L #NE# I: @FOR( OP(R)| @IN( JK, K, R):
43] @FOR( OP(J)| @IN( JK, K, J) #AND# @IN( PRED, K, J, R):
44] X( I, J) + X( L, R) - Y( I, L, J, K) < 1)))));
45]
46] @FOR( AS(K): @FOR( MA(I): @FOR( MA(L)| L #NE# I: @FOR( OP(R)| @IN( JK, K, R):
47] @FOR( OP(J)| @IN( JK, K, J) #AND# @IN( PRED, K, J, R):
48] - X( I, J) - X( L, R) + 2*Y( I, L, J, K) < 0)))));
49]
50] !RESTRICTING DECISION VARIABLES TO ZERO AND BINARY;
51] @FOR( MA(I): @FOR( MA(L)| L #EQ# I: @FOR( AS(K): @FOR( OP(J)| @IN( JK, K, J):
52] Y( I, L, J, K) = 0))));
53]
54] @FOR( MA(I): @FOR( MA(L)| L #NE# I: @FOR( AS(K):
55] @FOR( OP(J)| #NOT# @IN( JK, K, J) #OR# @IN( UL, K, J): Y( I, L, J, K) = 0))));
56]
57] @FOR( MA(I): @FOR( MA(L)| L #NE# I: @FOR( AS(K): @FOR( OP(J)| @IN(JK, K, J):
58] @BIN( Y( I, L, J, K))))));
59]
60] @FOR( ASS: @BIN( X));
61]
62] DATA:
63] S = @FILE(M3.LDT);
64] D = @FILE(M3.LDT);
65] P = @FILE(M3.LDT);
66] Q = @FILE(M3.LDT);
67] LAMBDA = ?;
68] ENDDATA
END
```

Fig. 3.3. LINGO input file for model **M3**

hour. LINGO software has been chosen since it permits a compact problem specification using a mathematical modeling language. An example of LINGO input file for model **M3** is shown in Fig. 3.3.

3.5 Simultaneous loading and routing

The mathematical models presented in this section aim at bi-objective optimization of station workloads and station-to-station product flows. The problem objective is to determine an assignment of tasks to stations and the corresponding assembly routes for all products so as to minimize the maximum workload of a station and the total transportation time between the stations.

The assignment of assembly tasks to stations simultaneously determines a set of alternative assembly routes for each product. Therefore, the FAS loading problem considered is actually a simultaneous station loading and product routing in which the corresponding decisions are made jointly.

3.5.1 Problem formulations

The models presented in this subsection can be classified into the following two categories ([106, 107, 110]):

1. Models **M7, M8** for fixed assembly routes, where each task can be assigned to only one station
2. Models **M9, M10** for alternative assembly routes, where each task can be assigned to more than one station.

The models are constructed for a general FAS, in which revisiting of stations is allowed and assembly process of each product type is subject to individual precedence constraints.

A feasible solution must satisfy the following three basic types of constraints:

– The demand for all product types must be satisfied.
– The subset of tasks assigned to each station must not exceed the station working space available.
– Products must be routed to the stations where the required tasks have been assigned, subject to precedence relations.

In order to achieve the FAS loading objectives one seeks to minimize the maximum station workloads, the maximum transfer time required to move products from one station to another or the total transportation time required to transfer all products between the stations.

The maximum workload P_{max}, the maximum transfer time to a station Q_{max}, and the total transportation time Q_{sum}, are defined below. (The notation used in the models is shown in Table 3.1.)

If only a single assembly route may be selected for each product type, the objective functions can be expressed as follows:

$$P_{max} = \max_{i \in I} \{ \sum_{k \in K} \sum_{j \in J_k} d_k p_{jk} x_{ij} \} \qquad (3.54)$$

$$Q_{max} = \max_{i \in I} \{ \sum_{k \in K} \sum_{l \neq i} \sum_{j \in J_k} d_k q_{li} y_{lijk} \} \qquad (3.55)$$

$$Q_{sum} = \sum_{i \in I} \sum_{l \neq i} \sum_{k \in K} \sum_{j \in J_k} d_k q_{il} y_{iljk} \qquad (3.56)$$

Otherwise, i.e., when more than one route may be selected for each product type, the objective functions are defined as below.

$$P_{max} = \max_{i \in I} \{ \sum_{k \in K} \sum_{j \in J_k} \sum_{l \in I} p_{jk} Y_{iljk} \} \qquad (3.57)$$

$$Q_{max} = \max_{i \in I} \{ \sum_{k \in K} \sum_{l \neq i} \sum_{j \in J_k} q_{li} Y_{lijk} \} \qquad (3.58)$$

$$Q_{sum} = \sum_{i \in I} \sum_{l \neq i} \sum_{k \in K} \sum_{j \in J_k} q_{il} Y_{iljk} \qquad (3.59)$$

A bicriterion machine loading and product routing problem can be formulated by selecting either P_{max} and Q_{max} or P_{max} and Q_{sum} as a pair of the objective functions. The corresponding models **M7**, **M8** and **M9**, **M10** are presented below.

Model M7. *Balancing station workloads and transfer times between the stations: fixed assembly routes*

Minimize

$$P_{max}, Q_{max} \qquad (3.60)$$

subject to

$$\sum_{k \in K} \sum_{j \in J_k} d_k p_{jk} x_{ij} \leq P_{max}; \quad i \in I \qquad (3.61)$$

$$\sum_{k \in K} \sum_{l \neq i} \sum_{j \in J_k} d_k q_{li} y_{lijk} \leq Q_{max}; \quad i \in I \qquad (3.62)$$

$$\sum_{i \in I_j} x_{ij} = 1; \quad j \in J \qquad (3.63)$$

$$\sum_{j \in J} a_{ij} x_{ij} \leq b_i; \quad i \in I \qquad (3.64)$$

$$x_{ij} + x_{lr} - y_{iljk} \leq 1; \quad k \in K, i \in I_j, l \in I_r, l \neq i, (j,r) \in R_k \qquad (3.65)$$

$$y_{iljk} \leq x_{ij}; \quad k \in K, i \in I_j, l \in I_r, l \neq i, (j,r) \in R_k \qquad (3.66)$$

$$y_{iljk} \leq x_{lr}; \quad k \in K, i \in I_j, l \in I_r, l \neq i, (j,r) \in R_k \qquad (3.67)$$

$$x_{ij} \in \{0,1\}; \quad \forall i,j \qquad (3.68)$$

$$y_{iljk} \geq 0; \quad \forall i,l,j,k \qquad (3.69)$$

58   3. Loading and Routing Decisions in Flexible Assembly Systems

The objective function P_{max} denotes total assembly time of the tasks assigned to the bottleneck station defined by constraint (3.61), and Q_{max} is the total transfer time required to deliver products to the station defined by constraint (3.62). Constraint (3.63) ensures that each task is assigned to only one station, and (3.64) is the station capacity constraint. Constraints (3.65), (3.66) and (3.67) ensure that each product type is routed to the stations where the required tasks are assigned. Implicitly, (3.65), (3.66) and (3.67) account for the individual precedence constraints of each product type.

Let us notice that for the fixed assembly routes (i.e., with task assignment constraint $\sum_{i \in I_j} x_{ij} = 1, \forall j \in J$), e.g., models **M3** and **M7**, part routing variable y_{iljk} represents product $x_{ij}x_{lr}$ of task assignment variables for each pair of immediate predecessor-successor tasks for product type $k \in K$, i.e., $y_{iljk} = x_{ij}x_{lr};\ k \in K,\ (j,r) \in R_k,\ i \in I_j,\ l \in I_r$. If both variables x_{ij} and x_{lr} are equal to 1, then $y_{iljk} = 1$, otherwise $y_{iljk} = 0$. It means that if task j is assigned to station i and the next task r $((j,r) \in R_k)$ to station l, then product type k after completion of task j on station i moves to station l to complete next task r.

The linear constraints (3.65), (3.66) and (3.67) defined for all $k \in K$, $(j,r) \in R_k$, $i \in I_j$, $l \in I_r$ transform into linear form the above quadratic relation between the routing and the task assignment variables (e.g.[40]).

Model **M7** has a special structure. The variables y_{iljk} need not to be binary constrained since the solution to the mixed integer program **M1** has 0-1 values of y_{iljk}.

Model M8. *Balancing station workloads and minimizing total transportation time: fixed assembly routes*

Minimize
$$P_{max},\ Q_{sum} \tag{3.70}$$

subject to (3.56), (3.61), (3.63), (3.64), (3.65), (3.66), (3.67), (3.68), (3.69).

Model M9. *Balancing station workloads and transfer times between the stations: alternative assembly routes*

Minimize
$$P_{max},\ Q_{max} \tag{3.71}$$

subject to

$$\sum_{i \in I_j}\sum_{l \in I_r} Y_{iljk} = d_k; \qquad k \in K,\ (j,r) \in R_k \tag{3.72}$$

$$\sum_{l \in I}(Y_{lijk} - Y_{ilrk}) = 0; \qquad i \in I,\ k \in K,\ (j,r) \in R_k \tag{3.73}$$

$$\sum_{k \in K}\sum_{j \in J_k}\sum_{l \in I} p_{jk}Y_{iljk} \leq P_{max}; \qquad i \in I \tag{3.74}$$

$$\sum_{k \in K} \sum_{j \in J_k} \sum_{l \neq i} q_{li} Y_{lijk} \leq Q_{max}; \qquad i \in I \qquad (3.75)$$

$$\sum_{i \in I_j} x_{ij} \geq 1; \qquad j \in J \qquad (3.76)$$

$$\sum_{j \in J} a_{ij} x_{ij} \leq b_i; \qquad i \in I \qquad (3.77)$$

$$Y_{iljk} \leq d_k x_{ij}; \qquad k \in K,\ i \in I_j,\ l \in I_r,\ (j,r) \in R_k \qquad (3.78)$$

$$Y_{iljk} \leq d_k x_{lr}; \qquad k \in K,\ i \in I_j,\ l \in I_r,\ (j,r) \in R_k \qquad (3.79)$$

$$x_{ij} \in \{0,1\}; \qquad \forall i,j \qquad (3.80)$$

$$Y_{iljk} \geq 0,\ \text{integer}; \qquad \forall i,l,j,k \qquad (3.81)$$

Constraint (3.72) ensures that all required tasks be allocated among the stations. Number of products k assigned to station i to perform task j is the sum of appropriate product outflows Y_{iljk} from station i after completion of task j, to all stations $l \in I_r$ capable of performing the next task r such that $(j,r) \in R_k$, i.e., $\sum_{l \in I_r} Y_{iljk}$. Equalities (3.73) are the network flow conservation equations for each station, product, and a pair of successively performed tasks. Constraints (3.74) and (3.75) define the maximum workload and the maximum transfer time. Constraint (3.76) ensures that each task is assigned to at least one station, and (3.77) is the station capacity constraint. Constraints (3.78) and (3.79) ensure that each product successively visits such stations where the required tasks may be assembled subject to precedence relations.

Model M10. *Balancing station workloads and minimizing total transportation time: alternative assembly routes*

Minimize

$$P_{max},\ Q_{sum} \qquad (3.82)$$

subject to (3.59), (3.72), (3.73), (3.74), (3.76), (3.77), (3.78), (3.79), (3.80), (3.81).

A common procedure to find efficient solutions to a multiobjective optimization problem which are not dominated by other solutions is to solve an equivalent single objective problem, where the original objective functions are combined into a single objective function using a set of weights.

The weighting method reduces the original bi-objective functions (3.60), (3.71) and (3.70), (3.82) into the following single objective functions (3.83) and (3.84), respectively.

$$\lambda P_{max} + (1-\lambda) Q_{max} \qquad (3.83)$$

$$\lambda P_{max} + (1-\lambda) Q_{sum} \qquad (3.84)$$

Let us denote the reduced problems **M7**, **M8**, **M9** and **M10**, respectively **M7**λ, **M8**λ, **M9**λ and **M10**λ.

The weight factor $\lambda \in [0,1]$ can be interpreted as a relative contribution of assembly and transportation times to the total completion time. However, such a contribution is not explicitly known beforehand. Therefore, a decision maker searches for a weight λ which would produce the most preferred solution. Obviously, the optimal solution to the **M7**λ, ..., **M10**λ for any particular $\lambda > 0$ would be an efficient solution to the corresponding bicriterion problem. A subset of strongly nondominated solutions can be generated by using the weights $0 < \lambda < 1$, otherwise weakly nondominated solutions are achieved.

3.5.2 An interactive heuristic for loading and routing

In this subsection an interactive approach is presented to solve the bicriterion problems **M7**, **M8**, **M9** and **M10** heuristically. The approach is based on searching for a weight λ to the objective functions which would produce a solution most preferred by the decision maker, e.g., [37]. At each iteration the decision maker further restricts the area of search in the set of all efficient solutions and guides the search along the efficient frontier for the most preferred solution. Briefly, the interactive approach proceeds as follows ([110]):

STEP 1. Set $t = 1$; t is the iteration number, $t = 1, 2, \ldots$
STEP 2. In the λ-range select 3 test weights λ_1, λ_2, λ_3, which allow for testing at one intermediate value plus the end points for each λ_p and $1 - \lambda_p$, $p = 1, 2, 3$.
STEP 3. Formulate 3 reduced single objective problems using the selected weights λ_p, $p = 1, 2, 3$.
STEP 4. Solve each of the reduced single objective problems using a suitable algorithm (optimal or a heuristic).
STEP 5. From the resulting solutions (task assignments and assembly routes) select the one most preferred based on the decision maker's preferences or a computer simulation of the corresponding detailed assembly schedules. For example, select the solution leading to the schedule with the shortest length.
STEP 6. In the λ-range under investigation, could there still be values of λ which might give a more preferred solution. If yes, set $t = t + 1$ and go to *STEP 2*. If no, stop.

In the first iteration, 3 test weights $\lambda_1 = \epsilon$, $\lambda_2 = 1 - \epsilon$ and $\lambda_3 = \frac{1}{2}$, uniformly distributed in the original λ-range $(0,1)$ are chosen, where ϵ is a sufficiently small positive number.

The search range decreases in the subsequent iterations. The search is concentrated in a region surrounding the weight λ_p which produced the preferred solution in the previous iteration. The new search region is obtained using a contraction mechanism in such a way that it contains the weight λ_p which produced the preferred solution. The new 3 test weights are chosen such that

they are uniformly distributed over the contracted λ-range.
The recursive relationships between weights λ_p^{t+1} and λ_p^t ($p = 1, 2, 3$) in the iterations $t+1$ and t are given below.

$$\lambda_p^{t+1} = \begin{cases} \frac{1}{2}\lambda_p^t & \text{if the preferred solution in iteration } t \text{ was obtained for } \lambda_p^t \in (0, \frac{1}{2}) \\ \frac{1}{2}(1 + \lambda_p^t) & \text{if the preferred solution in iteration } t \text{ was obtained for } \lambda_p^t \in (\frac{1}{2}, 1) \\ \lambda_p^t & \text{if the preferred solution in iteration } t \text{ was obtained for } \lambda_p^t = \frac{1}{2} \end{cases} \quad (3.85)$$

The procedure stops if there are no new solutions or when the search region becomes small enough for it to contain no new solutions which might be preferred by the decision maker.

In *STEP 4* either an optimal algorithm like a branch and bound or a heuristic can be used, especially for large-size problems. Using a heuristic does not guarantee that an optimal solution to the reduced single objective problem would be obtained. Instead, it is possible that a near-efficient solution to the bicriterion problem would be found which could be less preferred than a corresponding unobtained point. However, by solving several reduced single objective problems within the same λ-range, this efficient point can be uncovered even if a heuristic is used, e.g., [37, 110].

3.5.3 Numerical examples

In this subsection several numerical examples are presented to illustrate various applications of the models proposed. The models have been applied for loading of a hypothetical FAS made up of $m = 3$ identical stations, in which $n = 15$ task types are assembled to produce $v = 4$ product types. The corresponding ordered sequences of tasks $j \in J_k$ required to make each product type $k = 1, 2, 3, 4$ are the following: (1,2,3,4,6,12,14,15), (1,2,5,6,9,10,13,15), (2,4,5,7,8,9,10,14), (8,11,13,14,15). Each product type k has the same demand of $d_k = 100$ units.

The system is organized as a series of the stations 1,2,3 located along a bidirectional AGV guide path. The transfer time between the successive stations is equal to 2 time units, and hence transportation times q_{il} from station i to station l are: $q_{12} = q_{21} = 2$, $q_{13} = q_{31} = 4$, $q_{23} = q_{32} = 2$.

Assembly times p_{jk} for each task j are identical for all product types, (i.e., $p_{jk} = p_j, \forall k, \ j \in J_k$) and are as follows: $p_1 = 4$, $p_2 = 2$, $p_3 = 2$, $p_4 = 2$, $p_5 = 4$, $p_6 = 2$, $p_7 = 3$, $p_8 = 5$, $p_9 = 2$, $p_{10} = 4$, $p_{11} = 5$, $p_{12} = 2$, $p_{13} = 4$, $p_{14} = 2$, $p_{15} = 3$.

The components are assumed to be all of relatively similar sizes so that each part type feeder uses the same amount of a station finite working space. As a result, one can substitute $a_{ij} = 1$ for all i, j in the station capacity constraints (3.64), (3.77).

62   3. Loading and Routing Decisions in Flexible Assembly Systems

Fixed assembly routes. Models $\mathbf{M7}\lambda$ and $\mathbf{M8}\lambda$ with various values of the weight $\lambda \in \{0.00, 0.05, \ldots, 0.95, 1.00\}$ have been applied to determine task assignments and single assembly routes for a FAS, in which each station i ($i = 1, 2, 3$) has $b_i = 5$ part feeders.

Simple lower bounds on the maximum workload P_{max}, the maximum transfer time Q_{max}, and the total transfer time Q_{sum} for the example are

$$LBP_{max} = \lceil \sum_{k \in K} \sum_{J \in J_k} d_k p_{jk}/m \rceil = 2900 \quad (3.86)$$

$$LBQ_{max} = \min_{\Phi_1, \ldots, \Phi_m} \{\max_{i \in I}[(\lceil \frac{|\bigcup_{k \in \Phi_i} J_k|}{b_i}\rceil - 1) \min_{k \in \Phi_i}(d_k) \min_{l \neq i}(q_{il})]\} = 400 \quad (3.87)$$

$$LBQ_{sum} = \min_{\Phi_1, \ldots, \Phi_m} \{\sum_{i \in I}[(\lceil \frac{|\bigcup_{k \in \Phi_i} J_k|}{b_i}\rceil - 1) \min_{k \in \Phi_i}(d_k) \min_{l \neq i}(q_{il})]\} = 800 \quad (3.88)$$

where $\Phi_1, \ldots, \Phi_m$ denotes a partition of the set $\{1, \ldots v\}$ of v product types into m disjoint subsets, and $\lceil a \rceil$ is the smallest integer not less than a.

The three variants of the loading problem with fixed assembly routes have been considered in which the objective function accounts for: transfer times only ($\lambda = 0$), both assembly and transfer times ($0.05 \leq \lambda \leq 0.95$), and assembly times only ($\lambda = 1$). The solution results for a pair of criteria P_{max}, Q_{max} and P_{max}, Q_{sum} are shown in Tables 3.4 and 3.5, respectively. The only solution obtained for $\lambda \in [0.05, 0.85]$ is identical for both pairs of criteria.

For a comparison, problems $\mathbf{M7}\lambda$ and $\mathbf{M8}\lambda$ have also been solved for the case of unlimited station capacities, i.e., assuming that $b_i = n$ for all $i = 1, 2, 3$. For $\lambda \in [0.05, 0.85]$ the same values of the objective functions $P_{max} = 3000$, $Q_{max} = 800$ and $Q_{sum} = 1400$ have been obtained. Such an insensitivity to the problem parameters has not been observed for the other values of weight λ.

Notice that there exist alternate optima for the example, e.g., exchange of task assignments between stations 1 and 3 does not change values of the objective functions.

The integer programs $\mathbf{M7}\lambda$ and $\mathbf{M8}\lambda$ for the example are composed of 323 constraints with 195 variables.

Alternative assembly routes. The interactive procedure presented in Sect. 3.5.2 and models $\mathbf{M9}\lambda$, $\mathbf{M10}\lambda$ have been applied to workload balancing of a FAS with alternative assembly routes, where each station i ($i = 1, 2, 3$) has $b_i = 8$ part feeders. The lower bounds (3.87) and (3.88) on the maximum and the total transfer time are $LBQ_{max} = 200$ and $LBQ_{sum} = 200$, respectively.

In the first iteration, the optimal task assignments and assembly routes have been determined for weights $\lambda_1^1 = \epsilon = 0.05$, $\lambda_2^1 = 1 - \epsilon = 0.95$, $\lambda_3^1 = 0.50$.

3.5 Simultaneous loading and routing

Table 3.4. Task assignments and assembly routes for P_{max} and Q_{max} criteria: fixed routes

Task assignments			Assembly routes for products type:			
Station 1	Station 2	Station 3	1	2	3	4
(a) 1,2,3,4,7	5,6,8,9,10	11,12,13,14,15	1,2,3	1,2,3	1,2,1,2,3	2,3
(b) 1,2,3,4,5	6,7,8,9,10	11,12,13,14,15	1,2,3	1,2,3,1	1,2,3	2,3
(c) 1,2,4,5,7	3,6,8,11,13	9,10,12,14,15	1,2,1,2,3	1,2,3,2,3	1,2,3	2,3
(d) 1,11,12,13,14	2,3,7,8,10	4,5,6,9,15	1,2,3,1,3	1,2,3,2,1,3	2,3,2,3,2,1	2,1,3

(a) $\lambda = 0$, $P_{max} = 3400$, $Q_{max} = 800$
(b) $\lambda \in [0.05, 0.85]$, $P_{max} = 3000$, $Q_{max} = 800$ – the most preferred solution
(c) $\lambda \in [0.90, 0.95]$, $P_{max} = 2900$, $Q_{max} = 1000$
(d) $\lambda = 1$, $P_{max} = 2900$, $Q_{max} = 2000$

Table 3.5. Task assignments and assembly routes for P_{max} and Q_{sum} criteria: fixed routes

Task assignments			Assembly routes for products type:			
Station 1	Station 2	Station 3	1	2	3	4
(a) 1,2,3,4,5	6,7,8,11,12	9,10,13,14,15	1,2,3	1,2,3	1,2,3	2,3
(b) 1,2,3,4,5	6,7,8,9,10	11,12,13,14,15	1,2,3	1,2,3,1	1,2,3	2,3
(c) 1,2,4,5,7	3,8,9,10,11	6,12,13,14,15	1,2,1,3	1,3,2,3	1,2,3	2,3
(d) 1,3,8,9,11	2,5,6,7,13	4,10,12,14,15	1,2,1,3,2,3	1,2,1,3,2,3	2,3,2,1,3,1	1,2,3

(a) $\lambda = 0$, $P_{max} = 3500$, $Q_{sum} = 1400$
(b) $\lambda \in [0.05, 0.85]$, $P_{max} = 3000$, $Q_{sum} = 1400$ – the most preferred solution
(c) $\lambda \in [0.90, 0.95]$, $P_{max} = 2900$, $Q_{max} = 2200$
(d) $\lambda = 1$, $P_{max} = 2900$, $Q_{sum} = 3800$

For model **M9**λ, the solution for $\lambda_3^1 = 0.50$ with $P_{max} = 2900$ and $Q_{max} = 400$ is selected to be the most preferred solution. Therefore, the interactive search terminates at the first iteration, since $\lambda^{t+1} = \lambda^t$, see (3.85). For the best solution, the selected assembly routes are determined by the following set of flow variables Y_{iljk}:

$Y_{1,1,8,3} = 40$, $Y_{1,1,8,4} = 100$, $Y_{1,1,9,3} = 100$, $Y_{1,1,10,3} = 100$, $Y_{1,1,11,4} = 100$, $Y_{1,1,13,4} = 100$,
$Y_{1,1,14,4} = 100$, $Y_{2,1,7,3} = 40$, $Y_{2,1,8,3} = 60$, $Y_{2,2,6,2} = 100$, $Y_{2,2,7,3} = 60$, $Y_{2,2,9,2} = 100$,
$Y_{2,2,10,2} = 100$, $Y_{2,2,13,2} = 100$, $Y_{2,3,1,1} = 100$, $Y_{2,3,1,2} = 100$, $Y_{3,2,5,2} = 100$, $Y_{3,2,5,3} = 100$,
$Y_{3,3,2,1} = 100$, $Y_{3,3,2,2} = 100$, $Y_{3,3,2,3} = 100$, $Y_{3,3,3,1} = 100$, $Y_{3,3,4,1} = 100$, $Y_{3,3,4,3} = 100$,
$Y_{3,3,6,1} = 100$, $Y_{3,3,12,1} = 100$, $Y_{3,3,14,1} = 100$.

The assembly routes selected are shown in Fig. 3.4, where number of the station selected for each task to be performed is additionally indicated in parentheses.

For model **M10**λ, the most preferred nondominated solution with $P_{max} = 2900$ and $Q_{sum} = 460$ was obtained for $\lambda_2^1 = 0.95$, and hence in the second iteration **M10**λ was solved for three new weights (see (3.85)) $\lambda_1^2 = \frac{1}{2}(1+\lambda_1^1) = 0.525$, $\lambda_2^2 = \frac{1}{2}(1+\lambda_2^1) = 0.975$, $\lambda_3^2 = \frac{1}{2}(1+\lambda_3^1) = 0.750$. A new solution with $P_{max} = 2900$ and $Q_{sum} = 728$ was found only for $\lambda = 0.750$, whereas the

$k = 1, d_1 = 100 : 1(2) \to 2(3) \to 3(3) \to 4(3) \to 6(3) \to 12(3) \to 14(3) \to 15(3)$
$k = 2, d_2 = 100 : 1(2) \to 2(3) \to 5(3) \to 6(2) \to 9(2) \to 10(2) \to 13(2) \to 15(2)$
$k = 3, d_3 = 60 : 2(3) \to 4(3) \to 5(3) \to 7(2) \to 8(2) \to 9(1) \to 10(1) \to 14(1)$
$k = 3, d_3 = 40 : 2(3) \to 4(3) \to 5(3) \to 7(2) \to 8(1) \to 9(1) \to 10(1) \to 14(1)$
$k = 4, d_4 = 100 : 8(1) \to 11(1) \to 13(1) \to 14(1) \to 15(1)$

Fig. 3.4. Graph of assembly routes selected for model **M9**λ

other solutions are identical with those obtained in the first iteration. The new solution is dominated by the best solution found in the first iteration. Therefore, the search procedure stops and the optimal solution for $\lambda = 0.95$ is selected to be the best solution to the problem. The optimal assembly routes are defined by the following set of variables Y_{iljk}:

$Y_{1,1,2,3} = 100$, $Y_{1,1,4,3} = 100$, $Y_{1,1,5,3} = 100$, $Y_{1,1,7,3} = 100$, $Y_{1,1,8,3} = 100$, $Y_{1,1,9,3} = 100$,
$Y_{1,1,10,3} = 100$, $Y_{1,2,8,4} = 100$, $Y_{2,2,3,1} = 100$, $Y_{2,2,4,1} = 100$, $Y_{2,2,6,1} = 100$, $Y_{2,2,11,4} = 100$,
$Y_{2,2,12,1} = 100$, $Y_{2,2,13,2} = 29$, $Y_{2,2,13,4} = 100$, $Y_{2,2,14,1} = 100$, $Y_{2,2,14,4} = 99$, $Y_{2,3,14,4} = 1$,
$Y_{3,2,2,1} = 100$, $Y_{3,2,10,2} = 29$, $Y_{3,3,1,1} = 100$, $Y_{3,3,1,2} = 100$, $Y_{3,3,2,2} = 100$, $Y_{3,3,5,2} = 100$,
$Y_{3,3,6,2} = 100$, $Y_{3,3,9,2} = 100$, $Y_{3,3,10,2} = 71$, $Y_{3,3,13,2} = 71$.

The assembly routes selected are shown in Fig. 3.5, where number of the station selected for each task to be performed is additionally indicated in parentheses.

$k = 1, d_1 = 100 : 1(3) \to 2(3) \to 3(2) \to 4(2) \to 6(2) \to 12(2) \to 14(2) \to 15(2)$
$k = 2, d_2 = 71 : 1(3) \to 2(3) \to 5(3) \to 6(3) \to 9(3) \to 10(3) \to 13(3) \to 15(3)$
$k = 2, d_2 = 29 : 1(3) \to 2(3) \to 5(3) \to 6(3) \to 9(3) \to 10(3) \to 13(2) \to 15(2)$
$k = 3, d_3 = 100 : 2(1) \to 4(1) \to 5(1) \to 7(1) \to 8(1) \to 9(1) \to 10(1) \to 14(1)$
$k = 4, d_4 = 99 : 8(1) \to 11(2) \to 13(2) \to 14(2) \to 15(2)$
$k = 4, d_4 = 1 : 8(1) \to 11(2) \to 13(2) \to 14(2) \to 15(1)$

Fig. 3.5. Graph of assembly routes selected for model **M10**λ

The task assignments obtained for models **M9**λ and **M10**λ using the interactive procedure are shown in Tables 3.6 and 3.7, respectively. The results presented above for the best solutions indicate that model **M9**λ selected 2 alternative assembly routes for product type 3, whereas model **M10**λ selected 2 alternative routes for each product type 2 and 4. The assembly routes for the other solutions are not presented. The integer programs **M9**λ and **M10**λ for the example are composed of 564 constraints with 282 variables.

The effect of varying demand for products. The effect of varying demand for products under fixed and flexible routing was investigated assuming that demand d_1 for product type 1 increased from 100 to 200 units, all things being equal.

3.5 Simultaneous loading and routing   65

Table 3.6. Task assignments for P_{max} and Q_{max} criteria: alternative routes

λ	P_{max}, Q_{max}	Task assignments		
		Station 1	Station 2	Station 3
0.05	3250, 300	2,4,8,10,11,13,14,15	1,2,3,4,5,6,9,12	2,5,7,8,9,10,14,15
0.50	*2900, 400*	4,8,9,10,11,13,14,15	1,6,7,8,9,10,13,15	2,3,4,5,6,12,14,15
0.95	2900, 658	1,2,3,4,5,6,13,14	1,5,7,9,10,13,14,15	8,9,10,11,12,13,14,15
* * – the most preferred solution				

Table 3.7. Task assignments for P_{max} and Q_{sum} criteria: alternative routes

λ	P_{max}, Q_{sum}	Task assignments		
		Station 1	Station 2	Station 3
0.050	3300, 400	1,2,3,4,6,12,14,15	5,6,9,10,11,13,14,15	2,4,5,7,8,9,10,14
0.500, 0.525	3100, 400	1,2,5,6,9,10,13,15	3,4,6,11,12,13,14,15	2,4,5,7,8,9,10,14
0.750	2900, 728	1,2,3,4,5,6,7,8	4,6,9,10,12,13,14,15	7,8,9,10,11,13,14,15
0.950, 0.975	*2900, 460*	2,4,5,7,8,9,10,14	3,4,6,11,12,13,14,15	1,2,5,6,9,10,13,15
* * – the most preferred solution				

Tables 3.8 and 3.9 compare solution results obtained for a pair of criteria P_{max} and Q_{max}, for fixed and alternative assembly routes, respectively. The lower bounds (3.86) and (3.87) for the modified example data are $LBP_{max} = 3534$ and $LBQ_{max} = 400$.

For the fixed routing model $\mathbf{M7}\lambda$, the solution with $P_{max} = 3600$ and $Q_{max} = 1200$ obtained for $\lambda_2^1 = 0.95$ in the first iteration of the interactive heuristic dominates the other solutions, and hence in the second iteration $\mathbf{M7}\lambda$ was solved for weights $\lambda_1^2 = 0.525$, $\lambda_2^2 = 0.975$, $\lambda_3^2 = 0.750$. The only new solution with $P_{max} = 3700$ and $Q_{max} = 1000$ obtained for $\lambda = 0.750$ was selected to be the most preferred. In the third iteration $\mathbf{M7}\lambda$ was solved for $\lambda_1^3 = 0.7625$, $\lambda_2^3 = 0.9875$, $\lambda_3^3 = 0.875$, however the new solutions found are less preferred than that for $\lambda = 0.750$.

For the alternative routing model $\mathbf{M9}\lambda$, the solution with $P_{max} = 3534$ and $Q_{max} = 734$ obtained for $\lambda_2^1 = 0.95$ is selected in the first iteration. Then, in the second iteration $\mathbf{M9}\lambda$ was solved for three new weights $\lambda_1^2 = 0.525$, $\lambda_2^2 = 0.975$, $\lambda_3^2 = 0.750$. However, the only new solution obtained for $\lambda_2^2 = 0.975$ is dominated by that selected in the first iteration. For simplicity, the assembly routes selected are not presented.

Comparison of the most preferred solutions selected under fixed and flexible routing shows that if the demand for product type 1 is increased from 100 to 200 units, the maximum workload P_{max} will be increased from 3000 to 3700 for fixed routing and from 2900 to 3534 for flexible routing, whereas the maximum transfer time Q_{max} will be increased from 800 to 1000 and from 400 to 734, respectively for fixed and flexible routing.

The effect of varying transportation time. The effect of varying spacing and therefore transportation time was investigated assuming that the transfer time between successive stations is equal to the longest assembly time, that is to 5 time units, all things being equal. Hence, the transportation times q_{il} from station i to station l are: $q_{12} = q_{21} = 5$, $q_{13} = q_{31} = 10$, $q_{23} = q_{32} = 5$.

Table 3.8. Task assignments for $d_1 = 200$: fixed routes

λ	P_{max}, Q_{max}	Task assignments		
		Station 1	Station 2	Station 3
0.0500	4000, 1200	1,2,3,4,8	5,6,9,11,120	7,10,13,14,15
0.5000	3800, 1400	1,2,3,4,5	6,7,12,14,15	8,9,10,11,13
0.5250	3600, 1600	4,6,8,9,10	12,13,14,15	1,2,5,7,11
0.7500	*3700, 1000*	1,2,3,4,7	5,6,8,9,10	11,12,13,14,15
0.7625, 0.8750	3600, 1400	1,2,3,5,7	4,6,12,14,15	8,9,10,11,13
0.9500	3600, 1200	1,2,3,4,6	5,8,9,10,12	7,11,13,14,15
0.9750,0.9875	3600, 2000	1,2,3,5,12	4,6,7,14,15	8,9,10,11,13
* * – the most preferred solution				

Table 3.9. Task assignments for $d_1 = 200$: alternative routes

λ	P_{max}, Q_{max}	Task assignments		
		Station 1	Station 2	Station 3
0.050	4200,400	2,4,6,8,11,13,14,15	2,4,5,6,7,9,10,14	1,2,3,4,6,12,14,15
0.500,0.525	3700,800	2,3,4,6,10,12,14,15	1,2,4,6,8,9,13,14	2,4,5,6,7,11,14,15
0.750,0.950	*3534,734*	2,4,6,10,11,13,14,15	1,2,4,5,6,8,9,14	2,3,4,6,7,12,14,15
0.975	3534,800	2,4,5,6,7,8,14,15	1,2,4,6,9,11,13,14	2,3,4,6,10,12,14,15
* * – the most preferred solution				

Tables 3.10 and 3.11 compare solution results obtained for a pair of criteria P_{max} and Q_{sum}, for fixed and alternative assembly routes, respectively. The lower bound (3.88) for the modified example data is $LBQ_{sum} = 2000$.

For the fixed routing model **M8λ**, in the first iteration of the interactive heuristic the same solution with $P_{max} = 3000$ and $Q_{sum} = 3500$ was obtained for all three test weights $\lambda_1^1 = 0.05$, $\lambda_2^1 = 0.95$, $\lambda_3^1 = 0.50$. Therefore, the procedure terminates at the first iteration.

For the alternative routing model **M10λ**, the solution for $\lambda_3^1 = 0.50$ with $P_{max} = 3050$ and $Q_{sum} = 2050$ is selected to be the most preferred solution. Therefore, the interactive search also terminates at the first iteration, since $\lambda^{t+1} = \lambda^t$, see (3.85). The selected assembly routes are not presented.

Comparison of the most preferred solutions selected under fixed and flexible routing shows that if the transfer time between succesive stations increased from 2 to 5 time units, the maximum workload P_{max} will remain unchanged at 3000 for fixed routing and will increase from 2900 to 3050 for flexible routing, whereas the total transportation time Q_{sum} will increase from 1400 to 3500 and from 460 to 2050, respectively for fixed and flexible routing.

Table 3.10. Task assignments for $\min q_{il} = \max p_{jk} = 5$: fixed routes

λ	P_{max}, Q_{sum}	Task assignments		
		Station 1	Station 2	Station 3
0.05	3000, 3500	1,2,3,4,5	6,7,8,9,10	11,12,13,14,15
0.50	3000, 3500	1,2,3,4,5	6,7,8,9,10	11,12,13,14,15
0.95	3000, 3500	1,2,3,4,5	6,7,8,9,10	11,12,13,14,15

3.5 Simultaneous loading and routing 67

Table 3.11. Task assignments for min $q_{il} = \max p_{jk} = 5$: alternative routes

λ	P_{max}, Q_{sum}	Task assignments		
		Station 1	Station 2	Station 3
0.05	4500, 2000	1,2,4,5,6,7,14,15	2,8,9,10,11,13,14,15	2,3,4,5,6,12,14,15
0.50	*3050, 2050*	2,4,8,9,10,13,14,15	2,4,5,6,7,12,14,15	1,2,3,8,11,13,14,15
0.95	2900, 3000	2,4,5,7,8,11,14,15	2,4,5,9,10,13,14,15	1,2,3,4,6,12,14,15

* * – the most preferred solution

The above results have indicated that the flexible routing policy (for which the solution space is larger) is capable of better accommodating the varying process parameters and yields lower values of the objective functions.

Computational experiments. In order to evaluate the effectiveness of the approach proposed for FAS loading, 50 test problems have been solved using the interactive search procedure and models **M7λ**, **M8λ**, **M9λ**, **M10λ**. The test examples were constructed for a FAS with $3 \leq m \leq 5$ assembly stations each having $3 \leq b_i \leq 10$ part feeders all of similar sizes (i.e., all $a_{ij} = 1$), in which $10 \leq n \leq 40$ task types are required to simultaneously assembly $2 \leq v \leq 5$ product types, where each product type k requires $5 \leq |J_k| \leq 15$ different assembly tasks.

The assembly and transportation times p_{jk} and q_{il} were uniformly distributed over [1,10], and the demands d_k for product types were uniformly distributed over [25,100]. The interactive procedure was terminated if there were no new solutions found. For such a termination rule the number of iterations of the interactive heuristic for the test problems was not greater than 3.

The computational experiments have indicated that there exist many alternate optima that yield the same values of the objective functions for different task assignments, different assembly routes and weights λ. On the other hand, if alternative routes are allowed, the same task assignment with different assembly routes and different values of the objective functions can be obtained for different λ.

The maximum workload P_{max} more often reaches its lower bound LBP_{max} (3.86) for flexible routing as well as the other objectives functions take on lower values than those for fixed routing, all things being equal.

The experiments have also indicated that the flexible routing models **M9λ** and **M10λ** are much more sensitive to the varying relative importance between pairs of criteria, than models **M7λ** and **M8λ** for the fixed routing. If only single assembly routes are available, the optimal routes are slightly sensitive to the value of selected weight λ, while they are much more dependent on λ, when alternative routes are allowed.

In general, the tighter are station capacity constraints, and the closer values of assembly and transportation times, the more indifferent to the weight λ is the solution obtained using the interactive heuristic.

For many test examples ([110]) the weight λ of the best solution was observed to reflect the relative contribution of the two criteria to the value of the reduced single objective function obtained. The value of λ was found to be close to the ratio of $P_{max}/(P_{max}+Q_{max})$ and $P_{max}/(P_{max}+Q_{sum})$, respectively for (3.83) and (3.84). Such values of λ were observed particularly in the case of fixed routing.

The above results suggest some possible reduction of the computations required by the interactive search procedure. At each iteration $t \geq 2$ of the heuristic, instead of 3 reduced single objective problems formulated for 3 weights λ (3.85), one would need to solve only one reduced problem formulated for $\lambda^t = 1/(1 + Q^{t-1}/P^{t-1})$, where P^{t-1} and Q^{t-1} denote values of the two objective functions for the most preferred solution selected at iteration $t-1$. However, such a modification of the search procedure may further restrict the subset of efficient solutions generated using this approach.

The examples presented in this subsection and the test problems were solved using discrete optimizer LINGO [115]. CPU run times on a PC 486 have been varied from several seconds to over one hour.

The experiments have indicated that the computation time increases with:
- the number m of stations, n of task types and v of product types;
- the slackness of the system working space, e.g., measured by the ratio $(\sum_i^m b_i - n)/n$.

The CPU time required for solving **M9**λ and **M10**λ was much larger than that for **M7**λ and **M8**λ, and in particular **M10**λ required the largest computation time.

The routing variables representing station-to-station movements of products contribute essentially to the problem size, especially when alternative assembly routes are allowed. On the other hand, a typical FAS includes only a few quite versatile assembly stations, where only a subset of all required components is assembled simultaneously. For example, in mechanical assembly smaller subsets of parts are first assembled into several subassemblies which are next assembled into the final products. In such cases, the models proposed can be solved even by commercially available codes.

3.6 Sequential loading and routing

In this section an approximative sequential loading and routing procedure is proposed to determine task assignments and to select assembly routes for a mix of products so as to balance the station workloads and to minimize total transportation time. In the approach proposed, first the station workloads are balanced using a linear relaxation-based heuristic and then assembly routes are selected to minimize total transportation time based on a network flow model.

3.6.1 Problem formulations

A feasible solution of the combined loading and routing problem must satisfy the following four basic types of constraints:

- For each product all assembly tasks required must be completed.
- Each assembly task must be assigned to at least one station.
- The total space required for the tasks assigned to each station must not exceed the station finite working space available.
- Products must be successively routed to the stations where the required tasks have been assigned subject to precedence relations.

The following decision variables are used to model the loading and routing problem (for the other notation used, see Table 3.1):

x_{ij} = 1, if task j is assigned to station $i \in I_j$; otherwise $x_{ij} = 0$

y_{iljk} = 1, if product k is transferred from station i after completion of task j, to station l to perform next task; otherwise $y_{iljk} = 0$

z_{ijk} = 1, if product k is assigned to station i to perform task j; otherwise $z_{ijk} = 0$.

Let us notice that the product assignment variables z_{ijk} and the assembly routing variables y_{iljk} are dependent via flow balance equations

$$z_{ijk} = \sum_{l \in I_r} y_{iljk}; \quad k \in K, \ (j,r) \in R_k, \ i \in I_j \qquad (3.89)$$

The loading objective functions, i.e., the maximum workload P_{max}, and the total transportation time Q_{sum} are defined below.

$$P_{max} = \max_{i \in I} \{ \sum_{k \in K} \sum_{j \in J_k} \sum_{l \in I} p_{jk} y_{iljk} \} \qquad (3.90)$$

$$Q_{sum} = \sum_{i \in I} \sum_{l \neq i} \sum_{k \in K} \sum_{j \in J_k} q_{il} y_{iljk} \qquad (3.91)$$

The equation (3.90) can be simplified by replacing variables y_{iljk} with z_{ijk} using the relationship (3.89)

$$P_{max} = \max_{i \in I} \{ \sum_{k \in K} \sum_{j \in J_k} p_{jk} z_{ijk} \}$$

In the self explanatory short-hand notation used from now on to denote the different models presented in the sequel **L** denotes loading, **R** – routing, and **S** – selection of assembly plans.

A bicriterion machine loading and product routing problem has the following 0-1 programming formulation ([107]):

Model LR: *Balancing station workloads and minimizing total transportation time*

Minimize
$$P_{max}, Q_{sum} \tag{3.92}$$

subject to

$$\sum_{i \in I_j} \sum_{l \in I_r} y_{iljk} = 1; \qquad k \in K, (j,r) \in R_k \tag{3.93}$$

$$\sum_{l \in I} (y_{lijk} - y_{ilrk}) = 0; \qquad i \in I, k \in K, (j,r) \in R_k \tag{3.94}$$

$$\sum_{k \in K} \sum_{j \in J_k} \sum_{l \in I} p_{jk} y_{iljk} \leq P_{max}; \qquad i \in I \tag{3.95}$$

$$\sum_{i \in I} \sum_{l \neq i} \sum_{k \in K} \sum_{j \in J_k} q_{il} y_{iljk} = Q_{sum} \tag{3.96}$$

$$\sum_{i \in I_j} x_{ij} \geq 1; \qquad j \in J \tag{3.97}$$

$$\sum_{j \in J} a_{ij} x_{ij} \leq b_i; \qquad i \in I \tag{3.98}$$

$$y_{iljk} \leq x_{ij}; \qquad k \in K, i \in I_j, l \in I_r, (j,r) \in R_k \tag{3.99}$$

$$y_{iljk} \leq x_{lr}; \qquad k \in K, i \in I_j, l \in I_r, (j,r) \in R_k \tag{3.100}$$

$$x_{ij} \in \{0,1\}; \qquad \forall i,j \tag{3.101}$$

$$y_{iljk} \in \{0,1\}; \qquad \forall i,l,j,k \tag{3.102}$$

The first objective function in (3.92) represents imbalance of the workload distribution. Minimization of the maximum workload P_{max} subject to (3.95) implicitly equalizes the station workloads. Constraint (3.93) ensures for each product that all of its required tasks be allocated among the stations. Equalities (3.94) are the flow conservation equations for each station, product, and a pair of successively performed tasks. Constraints (3.95) and (3.96) define the workload of the bottleneck station and the total transportation time, respectively. Constraint (3.97) ensures that each task is assigned to at least one station, which admits alternative assembly routes for products. Constraint (3.98) is the station capacity constraint. Constraints (3.99) and (3.100) ensure that each product successively visits such stations where the required tasks may be assembled subject to precedence relations.

In order to find nondominated solutions to the bi-objective problem **LR** using the weighting approach, the following reduced single objective problem **LR**λ should be solved.

Model LRλ: *Minimizing weighted sum of the maximum workload and total transportation time*

Minimize
$$\lambda P_{max} + (1-\lambda) Q_{sum} \tag{3.103}$$

subject to (3.93) – (3.102).

A subset of nondominated solutions of **LR** can be found by solving **LRλ** for varying $\lambda \in [0,1]$.

Minimization of the weighted sum of assembly and transportation times for various weight λ may help to find task assignments and assembly routes of a schedule with the shortest length, see Chap. 7.

3.6.2 Lexicographic approach to loading and routing

An efficient solution to the bicriterion problem **LR** can be found by applying the lexicographic approach presented in this subsection, e.g., [107, 112]. In most practical situations the first objective of balancing the station workloads is more important for the FAS performance than the second one of minimizing the total transportation time. Hence, first one solves the **LR** problem with the objective of minimizing P_{max} and with constraint (3.96) omitted. Using equation (3.89) the problem can be reformulated into the following loading problem **L**.

Model L: *Balancing station workloads*

Minimize
$$P_{max} \tag{3.104}$$

subject to

$$\sum_{i \in I_j} z_{ijk} = 1; \quad k \in K, j \in J_k \tag{3.105}$$

$$\sum_{k \in K} \sum_{j \in J_k} p_{jk} z_{ijk} \leq P_{max}; \quad i \in I \tag{3.106}$$

$$\sum_{j \in J} a_{ij} x_{ij} \leq b_i; \quad i \in I \tag{3.107}$$

$$\sum_{i \in I_j} x_{ij} \geq 1; \quad j \in J \tag{3.108}$$

$$z_{ijk} \leq x_{ij}; \quad k \in K, j \in J_k, i \in I_j \tag{3.109}$$

$$x_{ij} = 0; \quad j \in J, i \notin I_j \tag{3.110}$$

$$z_{ijk} = 0; \quad j \in J, i \notin I_j, k \in K \tag{3.111}$$

$$x_{ij} \in \{0,1\}; \quad j \in J, i \in I_j \tag{3.112}$$

$$z_{ijk} \in \{0,1\}; \quad k \in K, j \in J_k, i \in I_j \tag{3.113}$$

72   3. Loading and Routing Decisions in Flexible Assembly Systems

The objective P_{max} (3.104) is a measure of system imbalance and represents workload of the bottleneck station defined by constraint (3.106). Equation (3.105) ensures that for each product all required tasks are allocated among the stations. Constraint (3.107) is the station capacity constraint. Constraint (3.109) ensures that each product is assigned to such stations where the required tasks may be assembled. Constraints (3.110) and (3.111) eliminate assignment of tasks and products to inappropriate stations.

Table 3.12. Objective functions

P_{max}	=	the maximum workload
$P_{max}^*, (P_{max}^H)$	=	the optimal (heuristic) solution value to problem **L**
P_{max}^R	=	the value of P_{max} obtained by solving problem **R** with P_{max}^H bound in (3.114)
Q_{sum}	=	the total transportation time
Q_{sum}^*	=	the optimal solution value to problem **LR**λ with $\lambda = 0$
Q_{sum}^R	=	the optimal solution value to problem **R** with P_{max}^H bound in (3.114)

Let P_{max}^* be the optimal value of P_{max} (3.104) (for notation used to differ various objective functions, see Table 3.12). Having solved problem **L**, the bi-objective problem **LR** is next reduced into the following single objective problem **L*R** of minimizing Q_{sum} for fixed $P_{max} = P_{max}^*$.

Model L*R: *Minimizing total transportation time for the balanced station workloads*

Minimize
$$Q_{sum} = \sum_{i \in I} \sum_{l \neq i} \sum_{k \in K} \sum_{j \in J_k} q_{il} y_{iljk}$$

subject to (3.93), (3.94), (3.97) – (3.102),

$$\sum_{k \in K} \sum_{j \in J_k} \sum_{l \in I} p_{jk} y_{iljk} \leq P_{max}^*; \quad i \in I \qquad (3.114)$$

The objective Q_{sum} is a measure of the material handling system total workload. Constraint (3.114) defines an upper bound on each station workload.

The reduced single objective problem is solved in the second step of the sequential approach. As a result, the task assignments x_{ij} and the assembly routes y_{iljk} are selected so as to to minimize total transportation time with the station workloads balanced at the lowest possible level.

An approximative solution to **L*R** can easily be found by solving the following simplified routing problem **R**, in which the task assignments x_{ij}^L and the maximum workload P_{max}^* obtained for the problem **L** remain fixed.

3.6 Sequential loading and routing

Model R: *Minimizing total transportation time for prefixed task assignments*

Minimize
$$Q_{sum} = \sum_{i \in I} \sum_{l \neq i} \sum_{k \in K} \sum_{j \in J_k} q_{il} y_{iljk}$$

subject to (3.93), (3.94), (3.102), (3.114),

$$y_{iljk} \leq x_{ij}^L x_{lr}^L; \ k \in K, i \in I_j, l \in I_r, (j,r) \in R_k \qquad (3.115)$$

In the above formulation constraint (3.115) is equivalent to variable upper bound constraints (3.99) and (3.100) for prefixed task assignments x_{ij}^L.

The routing problem has an embedded network flow structure and hence can be easily solved by using some specialized network optimization procedures (e.g., [6]). The routing problem can also be solved by direct application of an LP code and a simple rounding off procedure, if nonintegral solution is obtained or by solving Lagrangian relaxation of **R** with respect to the constraint (3.114).

3.6.3 A linear relaxation-based heuristic for loading

The loading problem **L** can be considered to be a generalized scheduling of unrelated parallel machines subject to additional resource constraint (3.107). If the station capacities were sufficiently large the above constraints could be neglected, and **L** would reduce to scheduling unrelated parallel machines. In [66] a simple rounding rule is shown which constructs an assignment on unrelated parallel machines with the maximum workload that is guaranteed to be no greater than twice the optimum.

This subsection presents a similar heuristic which finds a solution of the integer program **L** starting from the optimal solution of the LP relaxation of **L** (see [107, 112]).

The LP relaxation of **L** is formed by removing the integrality restrictions of x_{ij} and z_{ijk}. Let us call the resulting linear program LP(**L**). If one solves LP(**L**) with a simplex algorithm an optimal basic solution is found with a set of assignments $\tilde{x}_{ij}$ and $\tilde{z}_{ijk}$ for which some integrality constraint is violated. We call a task that is partially assigned to more than one station a *split task*. Each split task is associated with fractional assignment variables $\tilde{x}_{ij}$ $(0 < \tilde{x}_{ij} < 1)$ and $\tilde{z}_{ijk}$ $(0 < \tilde{z}_{ijk} < 1)$.

The heuristic proposed for the remaining assignment problem gives a "rounding" scheme for fractional assignments which leaves unchanged, as far as possible, the prefixed assignments which already satisfy integrality. The scheme is called $\delta - round \ off$ and proceeds as follows: Consider any basic solution to LP(**L**), any $0 < \delta < 1$ and each fractional task assignment $\tilde{x}_{ij}$ If $\tilde{x}_{ij} > \delta$, then assign task j to station i Otherwise, set $x_{ij} = 0$. The above rounding scheme can be summarized by the following formula

$$x_{ij} = \begin{cases} 1 & \text{if } \tilde{x}_{ij} > \delta \\ 0 & \text{if } \tilde{x}_{ij} \leq \delta \end{cases} \qquad (3.116)$$

The binary assignment variables x_{ij} obtained must satisfy the station capacity constraints, i.e.,

$$\sum_{j \in J_i} a_{ij} x_{ij} \leq b_i; \forall i$$

If all capacity constraints are satisfied, then product assignments z_{ijk} are determined based on $\tilde{z}_{ijk}$, using $\delta - round\ off$, and the final approximate solution of **L** is obtained.

If, however, the capacity constraint (3.107) for some station i is violated, i.e.,

$$\sum_{j \in J} a_{ij} x_{ij} > b_i,$$

then a violated valid inequality is found which is not satisfied by the solution $\tilde{x}$ of the linear programming relaxation of **L**.

The violated valid inequalities are generated in the following form (see [78])

$$\sum_{j \in C} x_{ij} \leq |C| - 1; \ i \in I \qquad (3.117)$$

where C is an unknown subset of J such that the following conditions hold for station i

$$\sum_{j \in C} a_{ij} > b_i \text{ and } \sum_{j \in C} \tilde{x}_{ij} > |C| - 1$$

In order to find a subset C let us introduce binary variables $\xi_j \in \{0,1\}; j \in J$ such that

$$\sum_{j \in J} a_{ij} \xi_j \geq b_i + 1 \text{ and } \sum_{j \in J} \tilde{x}_{ij} \xi_j > \sum_{j \in J} \xi_j - 1$$

The second inequality is equivalent to

$$\sum_{j \in J} (1 - \tilde{x}_{ij}) \xi_j < 1$$

Thus for each violated capacity constraint (3.107) we obtain the following knapsack separation problem for violated valid inequality (3.117)

$$\gamma_i = \min\{\sum_{j \in J}(1 - \tilde{x}_{ij})\xi_j : \sum_{j \in J} a_{ij}\xi_j \geq b_i + 1,\ \xi_j \in \{0,1\};\ j \in J\} \qquad (3.118)$$

If $\gamma_i < 1$, then subset C is determined by variables $\xi_j = 1$, i.e., $C = \{j : \xi_j = 1\}$.

The corresponding valid inequality (3.117) is violated by task assignment $\tilde{x}$ by the amount $(1 - \gamma_i)$.

In the linear relaxation-based algorithm proposed below violated valid inequalities generated in each iteration $t+1$ are added to the linear programming relaxation LP^t of **L** solved in the previous iteration t. The procedure is repeated until a feasible task assignment is found. Then, the corresponding product assignments z_{ijk} are determined by using $\delta - round\ off$ scheme

$$z_{ijk} = \begin{cases} 1 & \text{if } \tilde{z}_{ijk} > \delta \\ 0 & \text{if } \tilde{z}_{ijk} \leq \delta \end{cases} \quad (3.119)$$

and the algorithm terminates.

Linear Relaxation Heuristic for Loading

Input: The linear relaxation $LP(\mathbf{L})=LP^0$ of **L**.

Output: Heuristic solution: x_{ij}^H, z_{ijk}^H and P_{max}^H.

STEP 0.
Set $t = 0$, $LP^0 = LP(\mathbf{L})$,

STEP 1. Solution of the linear relaxation of the loading problem
Determine solution $\tilde{x}_{ij}$, $\tilde{z}_{ijk}$ and $\tilde{P}_{max}$ of the linear relaxation LP^t formulated in iteration t.

STEP 2. Task assignment
Determine task assignments x_{ij}^H applying $\delta-round\ off$ scheme (3.116). If the rounded off assignments satisfy capacity constraints (3.107) for all stations i, i.e.,

$$\sum_{j \in J} a_{ij} x_{ij}^H \leq b_i; \text{ for all } i \in I,$$

then go to *STEP 4*. Otherwise go to *STEP 3*.

STEP 3. Generation of violated valid inequalities
For each station i associated with a violated capacity constraint (3.107), find solution of the knapsack separation problem (3.118) ([69]) and construct the corresponding most violated valid inequality (3.117). Formulate the linear relaxation LP^t by adding the new violated valid inequalities to LP^{t-1}. Go to *STEP 1*.

STEP 4. Product assignment
Determine product assignments z_{ijk}^H applying $\delta - round\ off$ scheme (3.119), calculate the corresponding maximum workload
$P_{max}^H = max_{i \in I} \{\sum_{k \in K} \sum_{j \in J_k} p_{jk} z_{ijk}^H\}$ and terminate.

The above heuristic does not account for all violated valid inequalities of the loading problem **L**. Unlike strong cutting plane algorithms used for integer

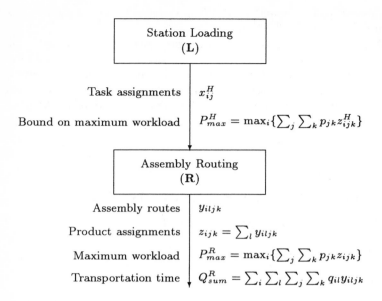

Fig. 3.6. A two-level task assignment and assembly routing

programming, the linear relaxation heuristic terminates, when $\delta - round\ off$ yields a feasible solution of **L**.

In most cases the maximum workload P_{max}^H determined by the above heuristic is greater than the optimum solution value P_{max}^* of the integer program **L**, i.e., $P_{max}^H > P_{max}^*$. Therefore P_{max} can be further improved at the second stage of the lexicographic approach while solving the routing problem **R** to select the best assembly routes (see Fig. 3.6). In the formulation of problem **R** the maximum workload P_{max} is now bounded from above by P_{max}^H (3.114) and task assignments x_{ij} are fixed at x_{ij}^H, whereas product assignments z_{ijk} remain free. Let us denote by $P_{max}^R = \max_{i \in I}\{\sum_{k \in K} \sum_{j \in J_k} p_{jk} z_{ijk}\}$ the resulting maximum workload and by Q_{sum}^R the corresponding solution value to **R**, see Table 3.12.

A further improvement of maximum workload can be achieved, for example, by exchanging task and product assignments between stations, e.g., [106]. For example, moving task j from station i to station i' may decrease maximum workload. Or, it may be that switching j and j' between stations i and i' decreases station workloads. Such 1-, 2-, and in general, s-exchanges (for arbitrary $s \leq m$) of the task assignments can be done until a locally optimal solution is reached. The above described exchanging procedure has been suggested in [106] to recover feasibility of a rounded off solution to the LP-relaxation of the loading problem, where the LP-based heuristic does not account for the violated valid inequalities.

3.6.4 Numerical examples

In this subsection 2 numerical examples are presented to illustrate application of the sequential approach and the linear relaxation heuristic for loading.

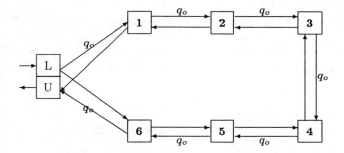

Fig. 3.7. FAS configuration for Example 1

Example 1. The FAS configuration for the example is provided in Fig. 3.7. The system is composed of $m = 6$ assembly stations ($i = 1, 2, 3, 4, 5, 6$) of various types, one loading/unloading station L/U ($i = 0/7$). The material handling system is bidirectional with $q_o = 2$ time units required for an AGV to move between any two neighbouring stations.

The production batch consists of $v = 4$ product types with the requirements 2,4,5, and 6 units of product type 1,2,3, and 4, respectively. All product types are assembled of $n = 8$ component types. The ordered sequences of components (assembly tasks) $j \in J_k$ required to make each product type $k = 1, 2, 3, 4$ are (4,6), (3,5,8), (1,2,4,7,8), (1,8), respectively.

$j_L \longrightarrow 4 \longrightarrow 6 \longrightarrow j_U$
$j_L \longrightarrow 3 \longrightarrow 5 \longrightarrow 8 \longrightarrow j_U$
$j_L \longrightarrow 1 \longrightarrow 2 \longrightarrow 4 \longrightarrow 7 \longrightarrow 8 \longrightarrow j_U$
$j_L \longrightarrow 1 \longrightarrow 8 \longrightarrow j_U$

Fig. 3.8. Graph of precedence relations

Graph of precedence relations is shown in Fig. 3.8 where j_L and j_U denote the beginning loading and the final unloading operations, respectively.

For the example problem, task assembly times depends upon product type and assembly station. For the same component type (task) j the assembly time p_{ijk} differs for different product types k and various stations i capable of performing j (for values of p_{ijk}, see Table 3.14). The loading and unloading times are identical for all products and are equal to 2 time units, i.e., $p_{L/U} = 2$.

78   3. Loading and Routing Decisions in Flexible Assembly Systems

For each assembly task j the working space a_{ij} required for the corresponding component feeders is independent on the station i, i.e., $a_{ij} = a_j$, $\forall i \in I_j$, $j \in J$ The values of a_j for various part type feeders are shown in the last row of Table 3.14.

The total working space b_i available at each station $i = 1,2,3,4,5,6$ is: $b_1 = 21$, $b_2 = 17$, $b_3 = 16$, $b_4 = 10$, $b_5 = 17$, $b_6 = 14$.

First, the bi-objective loading and routing problem is solved for a pair of criteria P_{max} and Q_{max} using the weighting approach. A subset of Pareto optimal solutions to **M9**λ is determined for $\lambda \in \{0.00, 0.05, \ldots, 0.95, 1.00\}$. The solution values of P_{max} and Q_{max} obtained for various λ are shown in Table 3.13.

Table 3.13. Solution results for model M9λ

λ	0.00	0.05	[0.10,0.20]	0.25	0.30	[0.35,0.60]	[0.65,0.95]	1.00
P_{max}	274	162	90	80	68	60	60	60
Q_{max}	34	34	42	52	54	60	64	72

Table 3.14. Assembly times and product assignments for model M9λ ($\lambda = 0.5$)

k	1			2			3				4	
j	4	6	3	5	8	1	2	4	7	8	1	8
$i=1$	2/2†	2/2				2/1	4		8	10/5		36
$i=2$	4	3				3/4	2/5		10	10		36/1
$i=3$	4	5			12	5	3	12	8/1	10		24/2
$i=4$			5	2	12			8	10		4	20/3
$i=5$			5	4	6/2			12	8/4		2/6	36
$i=6$			3/4	1/4	12/2			4/5	10		5	36
a_j	2	2	2	1	9	1	1	2	7	9	1	9
† p_{ijk}/Z_{ijk}												

A sample solution for **M9**λ with $\lambda = 0.5$ is shown below. The solution values are $P_{max} = 60$ and $Q_{max} = 60$ with the assembly routes determined by the following set of product flow variables:

- for product type 1
    $Y_{0,1,j_L,1} = 2;$ $Y_{1,1,4,1} = 2;$ $Y_{1,7,6,1} = 2$
- for product type 2
    $Y_{0,6,j_L,2} = 4;$ $Y_{6,6,3,2} = 4;$ $Y_{6,6,5,2} = 2;$ $Y_{6,5,5,2} = 2;$
    $Y_{5,7,8,2} = 2;$ $Y_{6,7,8,2} = 2$
- for product type 3
    $Y_{0,1,j_L,3} = 1;$ $Y_{0,2,j_L,3} = 4;$ $Y_{1,2,1,3} = 1;$ $Y_{2,2,1,3} = 4;$
    $Y_{2,6,2,3} = 5;$ $Y_{6,3,4,3} = 1;$ $Y_{6,5,4,3} = 4;$ $Y_{3,1,7,3} = 1;$
    $Y_{5,1,7,3} = 4;$ $Y_{5,7,8,3} = 5$
- for product type 4
    $Y_{0,5,j_L,4} = 6;$ $Y_{5,2,1,4} = 1;$ $Y_{5,3,1,4} = 2;$ $Y_{5,4,1,4} = 3;$
    $Y_{2,7,8,4} = 1;$ $Y_{3,7,8,4} = 2;$ $Y_{4,7,8,4} = 3.$

Table 3.14 shows the corresponding product assignments $Z_{ijk} = \sum_l Y_{iljk}$ for the example.

The bi-objective loading and routing problem is next solved for a pair of criteria P_{max} and Q_{sum} using the weighting approach and the lexicographic approach.

First, a subset of Pareto optimal solutions to **M10λ** is determined for $\lambda \in \{0.00, 0.05, \ldots, 0.95, 1.00\}$. The solution values of P_{max} and Q_{sum} obtained for various λ are shown in Table 3.15.

Table 3.15. Solution results for model **M10λ**

λ	0.00	0.05	[0.10,0.15]	[0.20,0.55]	[0.60,0.65]	[0.70,0.75]	0.80	[0.85,1.00]
P_{max}	310	187	115	76	70	66	68	60
Q_{sum}	108	120	128	144	160	180	180	202

In order to use the lexicographic approach and models **L** and **R**, the example data are modified appropriately to be compatible with all zero-one programming formulations of **L** and **R**. Each product type k with requirement d_k is replaced with new d_k product types, each with a unit requirement. The final solution results (product assignments and assembly routes) for the modified example data have been transformed back again to the original problem formulation.

First, problems **L** and **R** are solved sequentially. The solution values obtained are $P^L_{max} = 60$ and $Q^R_{sum} = 202$. For a comparison, the reduced single objective problem **M10λ** is solved for $\lambda = 1/(1+Q^R_{sum}/P^L_{max}) = 0.2290$. This value of λ reflects the relative contribution of the two criteria to the value of the reduced single objective function. The new solution values obtained are $P_{max} = 76$ and $Q_{sum} = 144$. The corresponding product assignments Z_{ijk} achieved are shown in Table 3.16.

Table 3.16. Assembly times and product assignments for model **M10λ** ($\lambda = 0.2290$)

k	1		2			3				4	
j	4	6	3	5	8	1	2	4	7	8	1	8
$i=1$	2/2†	2/2				2/5	4		8/1	10/1		36/1
$i=2$	4	3				3	2/5		10/2	10/4		36
$i=3$	4	5			12	5	3	12/5	8/2	10		24
$i=4$			5	2	12			8	10		4/3	20/3
$i=5$			5	4	6			12	8		2/2	36/2
$i=6$			3/4	1/4	12/4			4	10		5/1	36
† p_{ijk}/Z_{ijk}												

For a comparison, product assignments and assembly routes for the example problem have also been determined by using the sequential approach with the LP-based heuristic for loading. The solution procedure is as follows. First, an approximate solution of the loading problem **L** is determined by applying the linear relaxation-based heuristic with $\delta - round\ off$ scheme for $\delta = \frac{1}{2}$.

80   3. Loading and Routing Decisions in Flexible Assembly Systems

Table 3.17. Assembly times and product assignment for model **L** and the LP-based heuristic

k	1		2			3					4	
j	4	6	3	5	8	1	2	4	7	8	1	8
i=1	2/2†	2/2				2/5	4		8/3	10/2		36
i=2	4	3				3	2/5		10	10/3		36
i=3	4	5			12	5	3	12	8	10		24/2
i=4			5	2	12			8	10		4	20/3
i=5			5	4	6/4			12	8/2		2/6	36
i=6			3/4	1/4	12			4/5	10		5	36/1
† p_{ijk}/Z_{ijk}												

The resulting product assignments Z_{ijk} are shown in Table 3.17 at the corresponding assembly times p_{ijk}. The maximum workload is $P_{max}^H = 72$. Then, in the second stage of the sequential approach the final solution is determined by solving the routing problem **R**. The resulting total transportation time is $Q_{sum}^R = 208$, whereas the maximum workload remains the same $P_{max}^R = 72$.

The solution results have indicated that for models **M9λ** and **M10λ** there exist many alternate optima that yield the same values of the objective functions for different task assignments, different assembly routes and different weights λ.

Example 2. The next numerical example is presented to illustrate application of the sequential approach with the linear relaxation heuristic. Results of some computational experiments with the approach will be reported.

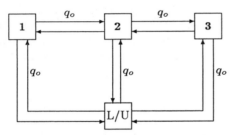

Fig. 3.9. FAS configuration for Example 2

The FAS configuration for the second example is provided in Fig. 3.9. The system is made up of $m = 3$ assembly stations ($i = 1, 2, 3$), and one loading/unloading station L/U. The material handling system is bidirectional with $q_o = 2$ time units required for an AGV to move between any two neighbouring stations.

The production batch consists of $v = 5$ selected products to be assembled of $n = 10$ types of components. The corresponding ordered sequences of tasks $j \in J_k$ required to assemble each product $k = 1, 2, 3, 4, 5$ are the following: (j_L,1,2,3,4,6,8,j_U), (j_L,1,2,4,5,6,7,9,10,j_U), (j_L,2,3,4,5,7,8,9,10,j_U), (j_L,1,3,5,6,7,8,9,10,j_U),

3.6 Sequential loading and routing   81

$j_L \longrightarrow 1 \longrightarrow 2 \longrightarrow 3 \longrightarrow 4 \longrightarrow 6 \longrightarrow 8 \longrightarrow j_U$
$j_L \longrightarrow 1 \longrightarrow 2 \longrightarrow 4 \longrightarrow 5 \longrightarrow 6 \longrightarrow 7 \longrightarrow 9 \longrightarrow 10 \longrightarrow j_U$
$j_L \longrightarrow 2 \longrightarrow 3 \longrightarrow 4 \longrightarrow 5 \longrightarrow 7 \longrightarrow 8 \longrightarrow 9 \longrightarrow 10 \longrightarrow j_U$
$j_L \longrightarrow 1 \longrightarrow 3 \longrightarrow 5 \longrightarrow 6 \longrightarrow 7 \longrightarrow 8 \longrightarrow 9 \longrightarrow 10 \longrightarrow j_U$
$j_L \longrightarrow 1 \longrightarrow 3 \longrightarrow 5 \longrightarrow 6 \longrightarrow 7 \longrightarrow 8 \longrightarrow 9 \longrightarrow 10 \longrightarrow j_U$

Fig. 3.10. Graph of precedence relations

(j_L,1,3,5,6,7,8,9,10,j_U), where j_L/j_U denotes loading/unloading operations (see Fig. 3.10).

The assembly times p_{jk} are shown below ($p_{jk} = 0$ indicates that task j is not required to assemble product k).

$$[p_{jk}] = \begin{bmatrix} 4,4,0,4,4 \\ 2,2,2,0,0 \\ 2,0,2,2,2 \\ 2,2,2,0,0 \\ 0,4,4,4,4 \\ 2,2,0,2,2 \\ 0,3,3,3,3 \\ 5,0,5,5,5 \\ 0,2,2,2,2 \\ 0,4,4,4,4 \end{bmatrix}$$

For each assembly task j the working space a_{ij} required for the corresponding component feeders is independent on the station i, i.e., $a_{ij} = a_j$, $\forall i \in I_j$, $j \in J$ ($a_{ij} = 0$ indicates that station i is incapable of performing task j).

$$[a_{ij}] = \begin{bmatrix} 1,2,3,1,2,3,0,0,0,5 \\ 0,0,0,1,2,3,1,2,3,5 \\ 1,2,3,0,0,0,1,2,3,0 \end{bmatrix}$$

The available working space is identical for all stations: $b_1 = 10$, $b_2 = 10$, $b_3 = 10$.

First, an approximate solution of the loading problem **L** has been determined applying the linear relaxation-based heuristic with δ − *round off* scheme for $\delta = \frac{1}{2}$. A feasible solution has been found after 4 iterations of the heuristic. Detailed results for each iteration are shown below.

Iteration 1. Solution of the linear programming relaxation of **L** yields the maximum workload $\tilde{P}_{max} = 38.667$ and the following nonzero task assignments $\tilde{x}_{ij}$:
$\tilde{x}_{1,1} = 1, \tilde{x}_{1,2} = 1, \tilde{x}_{1,3} = 1, \tilde{x}_{1,4} = 1, \tilde{x}_{1,5} = 1, \tilde{x}_{1,10} = 0.2$,
$\tilde{x}_{2,5} = 0.711, \tilde{x}_{2,6} = 1, \tilde{x}_{2,7} = 1, \tilde{x}_{2,8} = 0.158, \tilde{x}_{2,9} = 0.087, \tilde{x}_{2,10} = 0.800$,
$\tilde{x}_{3,1} = 1, \tilde{x}_{3,7} = 1, \tilde{x}_{3,8} = 0.842, \tilde{x}_{3,9} = 1$.
The δ − *round off* scheme (3.116) yields the following integral task assignments $x_{i,j}$:

82   3. Loading and Routing Decisions in Flexible Assembly Systems

$x_{1,1} = 1, x_{1,2} = 1, x_{1,3} = 1, x_{1,4} = 1, x_{1,5} = 1,$
$x_{2,5} = 1, x_{2,6} = 1, x_{2,7} = 1, x_{2,10} = 1,$
$x_{3,1} = 1, x_{3,7} = 1, x_{3,9} = 1.$
The task assignments violate capacity constraint (3.107) for station $i = 2$. Solution of the separation problem (3.118) for the violated constraint yields the following violated valid inequality (3.117)

$$x_{2,5} + x_{2,7} + x_{2,9} + x_{2,10} \leq 3$$

Iteration 2. Solution of the linear programming relaxation of **L** with additional constraint $x_{2,5} + x_{2,7} + x_{2,9} + x_{2,10} \leq 3$ yields:
$\tilde{x}_{1,1} = 1, \tilde{x}_{1,2} = 1, \tilde{x}_{1,3} = 1, \tilde{x}_{1,4} = 1, \tilde{x}_{1,5} = 1, \tilde{x}_{1,6} = 0.042,$
$\tilde{x}_{2,5} = 0.562, \tilde{x}_{2,6} = 0.958, \tilde{x}_{2,7} = 1, \tilde{x}_{2,10} = 1,$
$\tilde{x}_{3,1} = 1, \tilde{x}_{3,7} = 1, \tilde{x}_{3,8} = 1, \tilde{x}_{3,9} = 1.$
The maximum workload is $\tilde{P}_{max} = 38.667$.
The $\delta - round\ off$ scheme (3.116) yields the following task assignments:
$x_{1,1} = 1, x_{1,2} = 1, x_{1,3} = 1, x_{1,4} = 1, x_{1,5} = 1,$
$x_{2,5} = 1, x_{2,6} = 1, x_{2,7} = 1, x_{2,10} = 1,$
$x_{3,1} = 1, x_{3,7} = 1, x_{3,8} = 1, x_{3,9} = 1.$
The task assignments violate capacity constraint (3.107) again for station $i = 2$. Solution of the separation problem (3.118) for the violated constraint yields the following violated valid inequality (3.117)

$$x_{2,5} + x_{2,6} + x_{2,7} + x_{2,10} \leq 3$$

Iteration 3. Solution of the linear programming relaxation of **L** with two additional constraints $x_{2,5} + x_{2,7} + x_{2,9} + x_{2,10} \leq 3$, $x_{2,5} + x_{2,6} + x_{2,7} + x_{2,10} \leq 3$ yields:
$\tilde{x}_{1,1} = 1, \tilde{x}_{1,2} = 1, \tilde{x}_{1,3} = 1, \tilde{x}_{1,4} = 1, \tilde{x}_{1,10} = 0.600,$
$\tilde{x}_{2,5} = 1, \tilde{x}_{2,6} = 1, \tilde{x}_{2,7} = 0.600, \tilde{x}_{2,9} = 0.800, \tilde{x}_{2,10} = 0.400,$
$\tilde{x}_{3,1} = 1, \tilde{x}_{3,7} = 1, \tilde{x}_{3,8} = 1, \tilde{x}_{3,9} = 1.$
The maximum workload is $\tilde{P}_{max} = 38.667$.
The $\delta - round\ off$ scheme (3.116) yields the following task assignments:
$x_{1,1} = 1, x_{1,2} = 1, x_{1,3} = 1, x_{1,4} = 1, x_{1,10} = 1,$
$x_{2,5} = 1, x_{2,6} = 1, x_{2,7} = 1, x_{2,9} = 1,$
$x_{3,1} = 1, x_{3,7} = 1, x_{3,8} = 1, x_{3,9} = 1.$
The task assignments violate capacity constraint (3.107) for station $i = 1$. Solution of the separation problem (3.118) for the violated constraint yields the next violated valid inequality (3.117)

$$x_{1,2} + x_{1,3} + x_{1,4} + x_{1,10} \leq 3$$

Iteration 4. Solution of the linear programming relaxation of **L** with three additional constraints $x_{2,5} + x_{2,7} + x_{2,9} + x_{2,10} \leq 3$, $x_{2,5} + x_{2,6} + x_{2,7} + x_{2,10} \leq 3$, $x_{1,2} + x_{1,3} + x_{1,4} + x_{1,10} \leq 3$ yields:
$\tilde{x}_{1,1} = 1, \tilde{x}_{1,2} = 1, \tilde{x}_{1,3} = 1, \tilde{x}_{1,4} = 0.955, \tilde{x}_{1,5} = 0.367, \tilde{x}_{1,6} = 0.770,$
$\tilde{x}_{2,4} = 0.0450, \tilde{x}_{2,5} = 0.633, \tilde{x}_{2,6} = 0.367, \tilde{x}_{2,7} = 1, \tilde{x}_{2,8} = 0.243, \tilde{x}_{2,9} =$

$0.367, \tilde{x}_{2,10} = 1,$
$\tilde{x}_{3,1} = 1, \tilde{x}_{3,7} = 1, \tilde{x}_{3,8} = 0.757, \tilde{x}_{3,9} = 1.$
The maximum workload is $\tilde{P}_{max} = 38.667$.
The $\delta - round\ off$ scheme (3.116) yields the following task assignments:
$x_{1,1} = 1, x_{1,2} = 1, x_{1,3} = 1, x_{1,4} = 1, x_{1,6} = 1,$
$x_{2,5} = 1, x_{2,7} = 1, x_{2,10} = 1,$
$x_{3,1} = 1, x_{3,7} = 1, x_{3,8} = 1, x_{3,9} = 1.$
The above task assignments are feasible and satisfy capacity constraints (3.107) for all stations $i = 1, 2, 3$. The $\delta - round\ off$ scheme (3.119) yields the corresponding product assignments z_{ijk}:
$z_{1,1,1} = 1, z_{1,1,5} = 1, z_{1,2,1} = 1, z_{1,2,2} = 1, z_{1,2,3} = 1, z_{1,3,1} = 1, z_{1,3,3} = 1, z_{1,3,4} = 1, z_{1,3,5} = 1, z_{1,4,1} = 1, z_{1,4,2} = 1, z_{1,4,3} = 1, z_{1,6,1} = 1, z_{1,6,2} = 1, z_{1,6,4} = 1, z_{1,6,5} = 1,$
$z_{2,5,2} = 1, z_{2,5,3} = 1, z_{2,5,4} = 1, z_{2,5,5} = 1, z_{2,7,3} = 1, z_{2,10,2} = 1, z_{2,10,3} = 1, z_{2,10,4} = 1, z_{2,10,5} = 1,$
$z_{3,1,2} = 1, z_{3,1,4} = 1, z_{3,7,2} = 1, z_{3,7,4} = 1, z_{3,7,5} = 1, z_{3,8,1} = 1, z_{3,8,3} = 1, z_{3,8,4} = 1, z_{3,8,5} = 1, z_{3,9,2} = 1, z_{3,9,3} = 1, z_{3,9,4} = 1, z_{3,9,5} = 1.$
The maximum workload is $P^H_{max} = 45$.

Hence, iteration 4 of the linear relaxation heuristic yields an approximate solution of the loading problem **L**.

For a comparison, the exact solution of the integer program **L** for the example is shown below.

Maximum workload: $P^*_{max} = 39$

Task assignments:
$x_{1,1} = 1, x_{1,3} = 1, x_{1,4} = 1, x_{1,5} = 1, x_{1,6} = 1,$
$x_{2,8} = 1, x_{2,9} = 1, x_{2,10} = 1,$
$x_{3,1} = 1, x_{3,2} = 1, x_{3,7} = 1, x_{3,8} = 1, x_{3,9} = 1.$

Product assignments:
$z_{1,3,1} = 1, z_{1,3,3} = 1, z_{1,3,4} = 1, z_{1,3,5} = 1, z_{1,4,1} = 1, z_{1,4,2} = 1, z_{1,4,3} = 1, z_{1,5,2} = 1, z_{1,5,3} = 1, z_{1,5,4} = 1, z_{1,5,5} = 1, z_{1,6,1} = 1, z_{1,6,2} = 1, z_{1,6,4} = 1, z_{1,6,5} = 1,$
$z_{2,8,1} = 1, z_{2,8,4} = 1, z_{2,8,5} = 1, z_{2,9,2} = 1, z_{2,9,3} = 1, z_{2,9,4} = 1, z_{2,9,5} = 1, z_{2,10,2} = 1, z_{2,10,3} = 1, z_{2,10,4} = 1, z_{2,10,5} = 1,$
$z_{3,1,1} = 1, z_{3,1,2} = 1, z_{3,1,4} = 1, z_{3,1,5} = 1, z_{3,2,1} = 1, z_{3,2,2} = 1, z_{3,2,3} = 1, z_{3,7,2} = 1, z_{3,7,3} = 1, z_{3,7,4} = 1, z_{3,7,5} = 1, z_{3,8,3} = 1.$

Given the heuristic solution x^H_{ij}, P^H_{max} of the loading problem, the best assembly routes y_{iljk} (and the corresponding product assignments z_{ijk} (3.89)) are next determined in the second stage of the sequential approach by solving the routing problem **R**. Solution of **R** yields the maximum workload $P^R_{max} = 44$ and total transportation time $Q^R_{sum} = 40$. The assembly route selected for each product is shown in Fig. 3.11, where number of the station selected for each task is additionally indicated in parentheses.

Table 3.18 summarizes solution results for the example. It compares the results obtained using the sequential approach and the linear relaxation

$(L) \to 1(1) \to 2(1) \to 3(1) \to 4(1) \to 6(1) \to 8(3) \to (U)$
$(L) \to 1(1) \to 2(1) \to 4(1) \to 5(2) \to 6(1) \to 7(3) \to 9(3) \to 10(2) \to (U)$
$(L) \to 2(1) \to 3(1) \to 4(1) \to 5(2) \to 7(3) \to 8(3) \to 9(3) \to 10(2) \to (U)$
$(L) \to 1(1) \to 3(1) \to 5(2) \to 6(1) \to 7(3) \to 8(3) \to 9(3) \to 10(2) \to (U)$
$(L) \to 1(1) \to 3(1) \to 5(2) \to 6(1) \to 7(2) \to 8(3) \to 9(3) \to 10(2) \to (U)$

Fig. 3.11. Selected assembly routes

heuristic with those determined by solving bi-objective integer program **LR**λ for $\lambda \in \{0, 0.25, 0.50, 0.75, 1\}$ using discrete optimizer LINGO [115]. Table 3.18 shows that solution values for **LR**λ with $0.25 \leq \lambda \leq 0.75$ are independent on the selected values of the weight factor. However, the corresponding task assignments and assembly routes are not identical.

Table 3.18. Comparison of solution results: $b_1 = b_2 = b_3 = 10$

Objective function	Sequential approach LP(**L**), **R**	Bi-objective integer program **LR**λ				
		$\lambda = 0$	$\lambda = 0.25$	$\lambda = 0.50$	$\lambda = 0.75$	$\lambda = 1$
P_{max}	44	52	40	40	40	39
Q_{sum}	40	26	28	28	28	68

Table 3.19. Comparison of solution results: $b_1 = 10$, $b_2 = 9$, $b_3 = 8$

Objective function	Sequential approach LP(**L**), **R**	Bi-objective integer program **LR**λ				
		$\lambda = 0$	$\lambda = 0.25$	$\lambda = 0.50$	$\lambda = 0.75$	$\lambda = 1$
P_{max}	40	52	50	40	40	39
Q_{sum}	62	28	28	36	36	56

Table 3.20. Comparison of solution results: $b_1 = 15$, $b_2 = 12$, $b_3 = 10$

Objective function	Sequential approach LP(**L**), **R**	Bi-objective integer program **LR**λ				
		$\lambda = 0$	$\lambda = 0.25$	$\lambda = 0.50$	$\lambda = 0.75$	$\lambda = 1$
P_{max}	40	60	60	41	40	39
Q_{sum}	28	10	10	26	28	72

In order to show how the tightness of station capacity constraints (3.107) influence the solution, the above numerical example has also been solved with different values of parameters b_i, $i = 1, 2, 3$. The solution results are presented in Tables 3.19 and 3.20, respectively for the case with $b_1 = 10$, $b_2 = 9$, $b_3 = 8$ and $b_1 = 15$, $b_2 = 12$, $b_3 = 10$.

The results shown in Tables 3.18, 3.19 and 3.20 indicate that the tighter are station capacity constraints (3.107) the smaller is difference between the extreme values of single objective solutions to **LR**λ for $\lambda = 0, 1$. This type of information can be useful when achieving one objective alone at the sacrifice of the other objective.

3.6 Sequential loading and routing

In order to evaluate the effectiveness of the sequential loading and routing, 100 test problems have been solved using the LP-based heuristic with $\delta = \frac{1}{2}$. The test examples were constructed for a FAS with $3 \leq m \leq 6$ stations, in which $10 \leq n \leq 40$ tasks are required to simultaneously assemble $5 \leq v \leq 25$ products, where each product k requires $5 \leq |J_k| \leq 15$ different tasks to be performed. The assembly and transportation times p_{jk} and q_{il} were uniformly distributed over [1,10]. The routing problem **R** for the test examples was solved by using LINGO optimizer.

The solution values P_{max}^R have been compared with both the optimal solution value P_{max}^* of the loading problem **L** and the LP-based lower bound $\tilde{P}_{max}$ obtained in the last iteration of the linear relaxation heuristic for loading.

In order to evaluate the impact of the lexicographic approach on solution results, which significantly reduces the importance of the second objective function, the solution value Q_{sum}^R to the routing problem **R** has been compared with the solution value $Q_{sum}^{0.5}$ of problem **LRλ** for $\lambda = 0.5$.

In addition, the optimal solution value Q_{sum}^* of problem **LRλ** with $\lambda = 0$ has been compared with the solution value $Q_{sum}(P^*)$ corresponding to P_{max}^* and obtained by solving **LRλ** for $\lambda = 1$. The optimal solutions for the test problems were determined by using LINGO optimizer.

Table 3.21. Computational results

m, n, v	$\frac{P_{max}^R - \tilde{P}_{max}}{\tilde{P}_{max}}$	$\frac{P_{max}^R - P_{max}^*}{P_{max}^*}$	$\frac{Q_{sum}^R - Q_{sum}^{0.5}}{Q_{sum}^{0.5}}$	$\frac{Q_{sum}(P^*) - Q_{sum}^*}{Q_{sum}^*}$
3, 10, 5	11%	9%	47%	142%
3, 20, 15	9%	7%	32%	228%
4, 20, 15	7%	4%	36%	181%
4, 30, 20	11%	8%	81%	89%
5, 20, 15	14%	8%	67%	102%
5, 30, 20	10%	7%	8%	440%
5, 40, 25	16%	12%	28%	207%
6, 20, 15	8%	6%	51%	130%
6, 30, 20	15%	12%	19%	304%
6, 40, 25	16%	12%	74%	99%

P_{max}^R – the value of P_{max} obtained by using the sequential approach with the linear relaxation heuristic for loading
Q_{sum}^R – the optimal solution value to problem **R** with P_{max}^H bound in (3.114)
$\tilde{P}_{max}$ – LP-based lower bound on P_{max}
P_{max}^* – the optimal solution value to problem **L**
Q_{sum}^* – the optimal solution value to single objective problem **LRλ** with $\lambda = 0$
$Q_{sum}(P^*)$ – the value of Q_{sum} obtained by solving problem **LRλ** with $\lambda = 1$
$Q_{sum}^{0.5}$ – the value of Q_{sum} obtained by solving problem **LRλ** with $\lambda = 0.5$

The computational results are presented in Table 3.21. Ten problems were run for each set of parameters m, n, v which are shown in the first column of Table 3.21. For each set of ten test problems each entry in columns 2 and 3 represents, respectively the average percentage of solution value P_{max}^R over the lower bound $\tilde{P}_{max}$ and over the optimal solution value P_{max}^*. Each entry in columns 4 and 5 represents, respectively the average percentage of Q_{sum}^R over the solution value $Q_{sum}^{0.5}$ and the average percentage of $Q_{sum}(P^*)$ over Q_{sum}^*. That is, each entry in column 5 shows the relative difference between the extreme values of Q_{sum} obtained by solving **LR**λ for $\lambda = 1$ and $\lambda = 0$.

Table 3.21 indicates that the loading heuristic performs reasonably for the first objective of minimizing the maximum workload P_{max}. The average percentage over $\tilde{P}_{max}$ and over P_{max}^* is not greater than 16% and 12%, respectively.

The results obtained for the second objective function are interpreted somewhat different. The last two columns of Table 3.21 indicate that the greater is the relative difference between the maximum and the minimum solution values of Q_{sum} obtained for **LR**λ with $\lambda = 1$ and $\lambda = 0$, the closer is Q_{sum}^R to $Q_{sum}^{0.5}$. In most cases the solution value Q_{sum}^R to problem **R** is between $Q_{sum}^{0.5}$ and $Q_{sum}(P^*)$, that is between the extreme values of Q_{sum} obtained by solving **LR**λ for $\lambda \in [0.5, 1]$. The solution value of Q_{sum}^R is usually much better than the value $Q_{sum}(P^*)$ obtained for **LR**λ with $\lambda = 1$, where the objective function Q_{sum} is neglected.

For the test problems the CPU run time for the loading heuristic was not greater than several seconds on a PC 486.

4. Loading and Routing Decisions with Assembly Plan Selection

In this chapter a bicriterion machine loading and assembly routing with simultaneous assembly plan selection is considered for a general flexible assembly system and for a flexible assembly line.

Productivity of a flexible assembly system can be enhanced by allowing greater flexibility at the short-term planning. There are two sources that may increase the flexibility of the loading decision making: flexibility in the product assembly plans (i.e., precedence relations among the assembly tasks) and duplicate assignments of assembly tasks. In this chapter both the options are considered simultaneously. One can argue that fixing the sequence of assembly tasks for each product before the loading stage, without the knowledge of the task assignments and the product mix to be simultaneously assembled, decreases the chances of getting optimal or good workload balance. Avoiding premature selection of product assembly plans leads to a better balancing methodology. Therefore, in a good balancing procedure for a FAS, all duplicate task assignments and product assembly routes that simultaneously satisfy precedence relations among the tasks for all products, should be considered.

In the loading and routing problem considered in this chapter alternative assembly plans (i.e., alternative in-tree precedence relations among the assembly tasks) for each product are assumed to be available (e.g., [55]) as well as duplicate assignments of assembly task to different stations are allowed. Fig. 4.1 (similar to [21]) shows a simple practical example of a product made up of 5 components for which 10 different treelike assembly plans are available.

The problem considered is actually a combination of simultaneous machine loading, product routing and assembly plan selection. The problem is an extension of the FAS loading and routing problem presented in Chap. 3, where for each product type a single assembly plan was prefixed.

The modelling and solution approach proposed in this chapter is similar to that presented in Sect. 3.6. The loading, routing and assembly plan selection problem is formulated as a bicriterion 0-1 integer program. The objective is to simultaneously determine an allocation of assembly tasks among the stations and to select assembly plans and assembly routes for a mix of products so as to balance the station workloads and minimize total transportation time.

88   4. Loading and Routing Decisions with Assembly Plan Selection

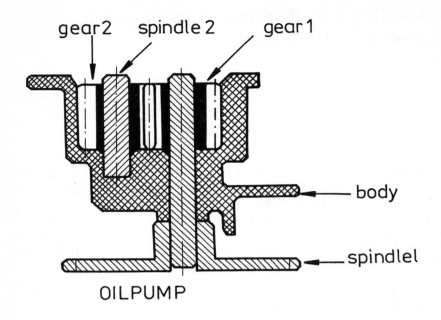

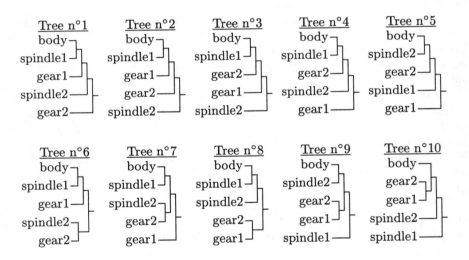

Fig. 4.1. An example of product and its alternative treelike assembly plans

4.1 Loading, routing and assembly plan selection in a general FAS

A two-level solution procedure is proposed. First, the station workloads are balanced using the linear relaxation-based heuristic presented in Sect. 3.6.3, and then the best assembly routes along with assembly plans are selected based on a network flow model to minimize total transportation time.

In Sect. 4.1 a general FAS is considered, where revisiting of stations is allowed, and a unidirectional flow line is modelled in Sect. 4.2.

4.1 Simultaneous loading, routing and assembly plan selection in a general flexible assembly system

In this section the bi-objective 0-1 integer programming formulation and a two-level solution approach is presented for simultaneous machine loading, product routing and assembly plan selection in a general flexible assembly system [111, 113].

Let us consider a FAS made up of m assembly stations $i \in I = \{1, \ldots, m\}$ and the L/U station connected by AGV paths. In the system n different types of assembly tasks $j \in J = \{1, \ldots, n\}$ can be performed to simultaneously assemble v products $k \in K = \{1, \ldots, v\}$ of various types. Let $I_j \subset I$ be the subset of stations capable of performing task j. Each station $i \in I$ has a finite working space b_i where a limited number of component feeders and gripper magazines can be placed. As a result only a limited number of assembly tasks can be assigned to one assembly station. Let a_{ij} be the amount of station $i \in I_j$ working space required for task j.

Each product k requires a subset J_k of assembly tasks to be performed subject to in-tree precedence relations defined by the assembly plan selected for this product. Let $S = \{1, \ldots, w\}$ be the index set of all assembly plans available and let T_s be the subset of tasks j in the plan $s \in S$. Each assembly plan $s \in S$ is represented by the set R_s of immediate predecessor-successor pairs of assembly tasks (j, r) such that task $j \in T_s$ must be performed immediately before task $r \in T_s$.

For each product k a subset $S_k \subset S$ of alternative assembly plans is available. The subsets S_k, $k \in K$ are disjoint so that each assembly plan $s \in S_k$ can be applied only for product k, i.e., for any two different products $k_1, k_2 \in K$, $k_1 \neq k_2$: $S_{k_1} \bigcap S_{k_2} = \emptyset$. Let us notice that $T_s = J_k$ for all assembly plans $s \in S_k$, $k \in K$.

Finally, denote by p_{jk} the assembly time required for task $j \in J_k$ of product k and by q_{il} the transportation time required to transfer a product from station i to station l.

The objective of the problem is to determine the optimal assignment of assembly tasks to stations and to select best assembly plans and assembly routes for all products so as to balance the station workloads and to minimize total transportation time.

A feasible solution of the combined loading, routing and assembly plan selection problem must satisfy the following five basic types of constraints:

- Each assembly task must be assigned to at least one station (duplicate assignments are allowed).
- For each product only one assembly plan must be selected.
- For each product and assembly plan selected all assembly tasks required must be completed.
- The total space required for the tasks assigned to each station must not exceed the station finite working space available.
- Each product must be successively routed to the stations where the required tasks have been assigned subject to precedence relations defined by the assembly plan selected.

4.1.1 Problem formulation

Table 4.1. Notation

		Indices
i	=	assembly station, $i \in I = \{1, \ldots, m\}$
j	=	assembly task, $j \in J = \{1, \ldots, n\}$
k	=	product, $k \in K = \{1, \ldots, v\}$
s	=	assembly plan, $s \in S = \{1, \ldots, w\}$
		Input parameters
a_{ij}	=	working space required for task j at station i
b_i	=	working space of station i
p_{jk}	=	assembly time for task j of product k
q_{il}	=	transportation time from station i to station l
I_j	=	the set of stations capable of performing task j
J_k	=	the set of tasks required for product k
R_s	=	the set of immediate predecessor-successor pairs of tasks (j, r) for assembly plan $s \in S$ such that task j must be performed immediately before task r
S_k	=	the set of assembly plans available for product k
T_s	=	the set of tasks in assembly plan s

In this subsection 0-1 integer programming formulation is presented for simultaneous loading, routing and assembly plan selection problem. To model the problem the following decision variables are introduced (for notation used, see Table 4.1):

u_s = 1, if assembly plan $s \in S$ is selected (for the corresponding product k such that $s \in S_k$); otherwise $u_s = 0$

x_{ij} = 1, if task j is assigned to station $i \in I_j$; otherwise $x_{ij} = 0$

y_{iljs} = 1, if for assembly plan s, after completion of task j the product is transferred from station i to station l to perform the next task; otherwise $y_{iljs} = 0$

z_{ijk} = 1, if product k is assigned to station i to perform task j; otherwise $z_{ijk} = 0$.

4.1 Loading, routing and assembly plan selection in a general FAS

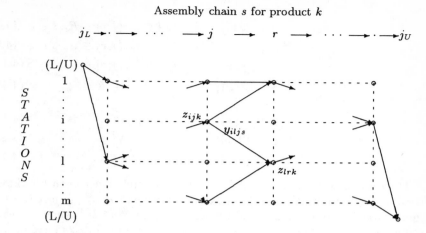

Fig. 4.2. Alternative assembly routes in a FAS

Let us notice that for each product $k \in K$, variables y_{iljs} (or z_{ijk}) must define a unique assembly route, since only one assembly plan $s \in S_k$ is selected for each $k \in K$ and subsets S_k are disjoint.

A bicriterion machine loading, product routing and assembly plan selection problem is formulated below as a bi-objective 0-1 integer program with a network flow structure embedded (see Fig. 4.2).

Model LRS: *Balancing station workloads and minimizing total transportation time in a general FAS*

Minimize
$$P_{max}, Q_{sum} \tag{4.1}$$

subject to

$$\sum_{i \in I_j} \sum_{l \in I} y_{iljs} = u_s; \qquad s \in S, j \in T_s \tag{4.2}$$

$$\sum_{l \in I} (y_{lijs} - y_{ilrs}) = 0; \qquad i \in I_r, (j,r) \in R_s, s \in S \tag{4.3}$$

$$\sum_{k \in K} \sum_{s \in S_k} \sum_{j \in J_k} \sum_{l \in I} p_{jk} y_{iljs} \leq P_{max}; \qquad i \in I \tag{4.4}$$

$$\sum_{i \in I} \sum_{l \neq i} \sum_{s \in S} \sum_{j \in T_s} q_{il} y_{iljs} = Q_{sum} \tag{4.5}$$

$$\sum_{i \in I_j} x_{ij} \geq 1; \qquad j \in J \tag{4.6}$$

$$\sum_{j \in J} a_{ij} x_{ij} \leq b_i; \qquad i \in I \qquad (4.7)$$

$$y_{iljs} \leq x_{ij}; \quad i \in I_j, l \in I_r, (j,r) \in R_s, s \in S \quad (4.8)$$

$$y_{iljs} \leq x_{lr}; \quad i \in I_j, l \in I_r, (j,r) \in R_s, s \in S \quad (4.9)$$

$$y_{iljs} \leq u_s; \quad i \in I_j, l \in I_r, (j,r) \in R_s, s \in S \quad (4.10)$$

$$\sum_{s \in S_k} u_s = 1; \qquad k \in K \qquad (4.11)$$

$$u_s \in \{0,1\}; \qquad \forall s \qquad (4.12)$$

$$x_{ij} \in \{0,1\}; \qquad \forall i,j \qquad (4.13)$$

$$y_{iljs} \in \{0,1\}; \qquad \forall i,l,j,s \qquad (4.14)$$

The first objective function in (4.1) represents imbalance of the workload distribution. Minimization of the maximum workload P_{max} subject to (4.4) implicitly equalizes the station workloads. Constraint (4.2) ensures for each product and assembly plan selected that all of its required tasks be allocated among the stations. Equalities (4.3) are the flow conservation equations for each station, assembly plan and a pair of successively performed tasks. Constraints (4.4) and (4.5) define the workload of the bottleneck station and the total transportation time, respectively. Constraint (4.6) ensures that each task is assigned to at least one station, and by this admits alternative assembly routes for products. Constraint (4.7) is the station capacity constraint. Constraints (4.8), (4.9) and (4.10) ensure that each product successively visits such stations where the required tasks may be assembled subject to precedence relations defined by the assembly plan selected. Constraint (4.11) ensures that only one assembly plan is selected for each product.

4.1.2 A two-level loading, routing and assembly plan selection

In this subsection a 2-level solution approach is presented based on the lexicographic approach and the LP-relaxation loading heuristic presented in Sect. 3.6.3.

Let us notice that the product assignment variables z_{ijk} and the assembly routing variables y_{iljs} are dependent via flow balance equations

$$z_{ijk} = \sum_{l \in I} \sum_{s \in S_k} y_{iljs}; \quad k \in K, \ j \in J_k, \ i \in I_j \qquad (4.15)$$

where, in view of (4.11), $\sum_{s \in S_k} y_{iljs} = y_{iljs'}$ for an assembly plan $s' \in S_k$, $u_{s'} = 1$, selected for product k.

Using relation (4.15) the single objective problem of minimizing P_{max} can be reformulated into the loading problem **L** presented in Sect. 3.6.2. The loading problem **L** can be solved by using the linear relaxation-based heuristic presented in Sect. 3.6.3.

4.1 Loading, routing and assembly plan selection in a general FAS

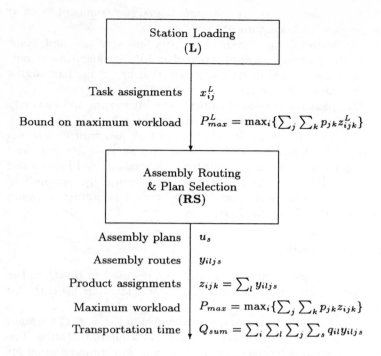

Fig. 4.3. A two-level task assignment, assembly routing and plan selection

Having solved problem **L**, the bi-objective problem **LRS** is next reduced into the following single objective simplified assembly routing and plan selection problem **RS** of minimizing Q_{sum} for prefixed task assignments x_{ij}^L obtained for **L** and with P_{max} bounded from above by P_{max}^L.

Model RS: *Minimizing total transportation time for prefixed task assignments*

Minimize
$$Q_{sum} = \sum_{i \in I} \sum_{l \neq i} \sum_{s \in S} \sum_{j \in J_s} q_{il} y_{iljs} \qquad (4.16)$$

subject to (4.2), (4.3), (4.11), (4.12), (4.14)

$$\sum_{k \in K} \sum_{s \in S_k} \sum_{j \in J_k} \sum_{l \in I} p_{jk} y_{iljs} \leq P_{max}^L; \qquad i \in I \qquad (4.17)$$

$$y_{iljs} \leq x_{ij}^L x_{lr}^L u_s; \quad i \in I_j, l \in I_r, (j,r) \in R_s, s \in S \quad (4.18)$$

The objective Q_{sum} is a measure of the material handling system total workload. Constraint (4.17) defines an upper bound on each station workload.

Constraint (4.18) is equivalent to variable upper bound constraints (4.8), (4.9) and (4.10) for prefixed task assignments x_{ij}^L.

The routing problem has an embedded network flow structure and hence can be easily solved, e.g., by direct application of an LP code and some rounding off procedure, if nonintegral solution is obtained or by solving Lagrangian relaxation of **RS** with respect to the constraint (4.17).

The two-level approach for station loading, assembly routing and assembly plan selection problem in a FAS is summarized in Fig. 4.3.

It should be pointed out that, in addition to task assignments x_{ij}, solution to the upper level problem **L** determines also an assembly plan and the corresponding assembly route selected for each product, and hence total transportation time. However, the two sets of variables u_s, z_{ijk} remain free for the second step of the solution procedure and their final improved values are determined only by solving the lower level problem **RS**.

4.1.3 Numerical examples

In this subsection a simple numerical example is presented to illustrate the application of the 2-level approach with the linear relaxation heuristic for loading.

The FAS configuration for the example is provided in Fig. 3.9. The system is made up of $m = 3$ assembly stations ($i = 1, 2, 3$), and one L/U station. The material handling system is bidirectional with $q_o = 2$ time units required for an AGV to move between any two neighbouring stations.

The production batch consists of $v = 5$ products to be assembled of $n = 10$ types of components using $w = 10$ assembly plans where two alternative assembly plans are available for each product. The assembly plans for products are in the form of assembly sequences (chains of tasks) to be performed. The available sequences $s \in S_k$ of tasks $j \in J_k$ required to assemble each product $k = 1, 2, 3, 4, 5$ are the following:

$$k = 1, \quad s = 1: \quad (j_L, 1, 2, 3, 4, 6, 8, j_U)$$
$$k = 1, \quad s = 2: \quad (j_L, 1, 2, 4, 3, 6, 8, j_U)$$
$$k = 2, \quad s = 3: \quad (j_L, 1, 2, 4, 5, 6, 7, 9, 10, j_U)$$
$$k = 2, \quad s = 4: \quad (j_L, 1, 2, 6, 4, 5, 7, 9, 10, j_U)$$
$$k = 3, \quad s = 5: \quad (j_L, 2, 3, 4, 5, 7, 8, 9, 10, j_U)$$
$$k = 3, \quad s = 6: \quad (j_L, 2, 7, 3, 4, 5, 8, 9, 10, j_U)$$
$$k = 4, \quad s = 7: \quad (j_L, 1, 3, 5, 6, 7, 8, 9, 10, j_U)$$
$$k = 4, \quad s = 8: \quad (j_L, 1, 3, 8, 5, 6, 7, 9, 10, j_U)$$
$$k = 5, \quad s = 9: \quad (j_L, 1, 3, 5, 6, 7, 8, 9, 10, j_U)$$
$$k = 5, \quad s = 10: \quad (j_L, 1, 3, 8, 5, 6, 7, 9, 10, j_U)$$

where j_L/j_U denotes loading/unloading operations.

Let us notice that products $k = 4, 5$ are of the same type, and hence the corresponding pairs of assembly sequences $s = 7, 9$ and $s = 8, 10$ are identical.

The assembly times p_{jk} are shown below ($p_{jk} = 0$ indicates that task j is not required to assemble product k).

$$[p_{jk}] = \begin{bmatrix} 4,4,0,4,4 \\ 2,2,2,0,0 \\ 2,0,2,2,2 \\ 2,2,2,0,0 \\ 0,4,4,4,4 \\ 2,2,0,2,2 \\ 0,3,3,3,3 \\ 5,0,5,5,5 \\ 0,2,2,2,2 \\ 0,4,4,4,4 \end{bmatrix}$$

For each assembly task j, the working space a_{ij} required for the corresponding component feeders is independent on the station i, i.e., $a_{ij} = a_j$, $\forall i \in I_j$, $j \in J$ ($a_{ij} = 0$ indicates that station i is incapable of performing task j).

$$[a_{ij}] = \begin{bmatrix} 1,2,3,1,2,3,0,0,0,5 \\ 0,0,0,1,2,3,1,2,3,5 \\ 1,2,3,0,0,0,1,2,3,0 \end{bmatrix}$$

The available working space for each station is: $b_1 = 10$, $b_2 = 9$, $b_3 = 8$.

First, an approximate solution of the loading problem **L** has been determined applying the linear relaxation-based heuristic A feasible solution has been found after 4 iterations of the heuristic. The results obtained are shown in Table 4.2.

Table 4.2. Solution of the loading problem

Task assignments		
Station 1	Station 2	Station 3
1,2,4,5,6	4,7,8,10	1,3,7,9
Maximum workload: $P_{max}^L = 40$		

Next, given the above solution x_{ij}^L, P_{max}^L of the loading problem, the best assembly plans u_s and assembly routes y_{iljs} (and the corresponding product assignments z_{ijk} (4.15)) have been determined by solving the routing problem **RS**. Direct application of an LP code to LP relaxation of **RS** yields an integer solution with the maximum workload $P_{max} = 40$ and total transportation time $Q_{sum} = 50$. The assembly sequence and route selected for each product are shown in Fig. 4.4, where the number of the station selected for each task to be performed is additionally indicated in parentheses.

96   4. Loading and Routing Decisions with Assembly Plan Selection

$k=1, s=1:$ $(L) \to 1(1) \to 2(1) \to 3(3) \to 4(2) \to 6(1) \to 8(2) \to (U)$
$k=2, s=3:$ $(L) \to 1(1) \to 2(1) \to 4(1) \to 5(1) \to 6(1) \to 7(3) \to 9(3) \to 10(2) \to (U)$
$k=3, s=6:$ $(L) \to 2(1) \to 7(3) \to 3(3) \to 4(2) \to 5(1) \to 8(2) \to 9(3) \to 10(2) \to (U)$
$k=4, s=8:$ $(L) \to 1(3) \to 3(3) \to 8(2) \to 5(1) \to 6(1) \to 7(3) \to 9(3) \to 10(2) \to (U)$
$k=5, s=10:$ $(L) \to 1(3) \to 3(3) \to 8(2) \to 5(1) \to 6(1) \to 7(3) \to 9(3) \to 10(2) \to (U)$

Fig. 4.4. Graph of assembly sequences and routes selected

Table 4.3 summarizes solution results for the example. It compares the results obtained using the 2-level approach and the linear relaxation heuristic with those determined by solving integer program **LRS** with reduced single objective function $\lambda P_{max} + (1-\lambda)Q_{sum}$ for $\lambda \in \{0, 0.5, 1\}$. Let us denote the resulting single objective integer program as **LRS**λ. The **LRS**λ was solved using discrete optimizer LINGO [115]. CPU time for the LINGO optimizer and the example was over two hours on a PC 486, whereas the computation time required for the 2-level approach with the linear relaxation heuristic was not greater than several seconds.

Table 4.3. Comparison of solution results

Objective function	2-level approach LP(L), RS	Integer program **LRS**λ with single objective: $\lambda P_{max} + (1-\lambda)Q_{sum}$		
		$\lambda = 0$	$\lambda = 0.5$	$\lambda = 1$
P_{max}	40	52	40	39
Q_{sum}	50	28	36	50
Selected sequences $s=$	1,3,6,8,10	1,3,5,7,9	1,3,5,7,9	1,3,5,7,9

Computational results In order to evaluate average performance of the 2-level approach with the linear relaxation loading heuristic 100 test problems were solved. The computational results are presented in Table 4.4. Ten problems were run for each set of parameters m, n, v, w which are shown in the first column of Table 4.4. For each set of 10 test problems each entry in columns 2 and 3 represents, respectively the average percentage of solution value P_{max}^{RS} over the lower bound $\tilde{P}_{max}$ and over the optimal solution value P_{max}^*. Each entry in columns 4 and 5 represents, respectively the average percentage of Q_{sum}^{RS} over the solution value $Q_{sum}^{0.5}$ and the average percentage of $Q_{sum}(P^*)$ over Q_{sum}^*. That is, each entry in column 5 shows the relative difference between the extreme values of Q_{sum} obtained by solving **LRS**λ for $\lambda = 1$ and $\lambda = 0$.

Table 4.4 indicates that the loading heuristic performs reasonably for the first objective of minimizing the maximum workload P_{max}. The average percentage over $\tilde{P}_{max}$ and over P_{max}^* is not greater than 15% and 11%, respectively.

The results obtained for the second objective function are interpreted somewhat different. The last two columns of Table 4.4 indicate that the greater is the relative difference between the maximum and the minimum

Table 4.4. Computational results

m, n, v, w	$\frac{P_{max}^{RS} - \tilde{P}_{max}}{\tilde{P}_{max}}$	$\frac{P_{max}^{RS} - P_{max}^*}{P_{max}^*}$	$\frac{Q_{sum}^{RS} - Q_{sum}^{0.5}}{Q_{sum}^{0.5}}$	$\frac{Q_{sum}(P^*) - Q_{sum}^*}{Q_{sum}^*}$
3, 10, 5, 10	8%	5%	27%	132%
3, 20, 15, 20	9%	6%	32%	118%
4, 20, 15, 25	6%	4%	26%	141%
4, 30, 20, 35	10%	7%	18%	89%
5, 20, 15, 20	12%	7%	17%	102%
5, 30, 20, 30	11%	9%	12%	340%
5, 40, 25, 40	15%	10%	23%	207%
6, 20, 15, 25	13%	9%	15%	330%
6, 30, 20, 35	12%	9%	10%	404%
6, 40, 25, 50	15%	11%	24%	99%

P_{max}^{RS} – the value of P_{max} obtained by using the two-level approach
Q_{sum}^{RS} – the optimal solution value to problem **RS**
$\tilde{P}_{max}$ – LP-based lower bound on P_{max}
P_{max}^* – the optimal solution value to problem **L**
Q_{sum}^* – the optimal solution value to problem **LRS**λ with $\lambda = 0$
$Q_{sum}(P^*)$ – the value of Q_{sum} obtained by solving problem **LRS**λ with $\lambda = 1$
$Q_{sum}^{0.5}$ – the value of Q_{sum} obtained by solving problem **LRS**λ with $\lambda = 0.5$

solution values of Q_{sum} obtained for **LRS**λ with $\lambda = 1$ and $\lambda = 0$, the closer is Q_{sum}^{RS} to $Q_{sum}^{0.5}$. In most cases the solution value Q_{sum}^{RS} to problem **RS** is between $Q_{sum}^{0.5}$ and $Q_{sum}(P^*)$, that is between the extreme values of Q_{sum} obtained by solving **LRS**λ for $\lambda \in [0.5, 1]$. The solution value of Q_{sum}^{RS} is usually much better than the value $Q_{sum}(P^*)$ obtained for **LRS**λ with $\lambda = 1$, where the objective function Q_{sum} is neglected.

For the test problems the CPU run time for the 2-level approach with the loading heuristic was not greater than several seconds on a PC 486.

4.2 Simultaneous loading, routing and assembly plan selection in a flexible assembly line

In this section the bicriterion 0-1 integer programming formulation and a two-level solution approach is presented for simultaneous machine loading, product routing and assembly plan selection in a flexible assembly line [109]. The problem is a restriction of the loading, routing and assembly plan selection problem presented in Sect. 4.1 for a general flexible assembly system, where multidirectional product flows and revisiting of stations were allowed.

A flexible assembly line (FAL) is a unidirectional flow system made up of a set of assembly stations in series and a loading/unloading (L/U) station, linked with an automated material handling system. A typical assembly process in a FAL proceeds as follows. A base part of a product is loaded on a pallet and enters the line at the L/U station. As the pallet is carried by a conveyor or an automated guided vehicle through a series of assembly stations, components are assembled with the base part. A product may bypass some stations but does not revisit any station. When all the required components are assembled with the base part, it is carried back to the L/U station and the complete product leaves the system.

When balancing the FAL all duplicate task assignments and product assembly routes that simultaneously satisfy the assembly precedence relations for all products in a unidirectional flow line should be considered.

Let us consider a FAL made up of m assembly stations $i \in I = \{1, \ldots, m\}$ in series $1, \ldots, m$ connected by a unidirectional AGV path, and the L/U station. In the system n different types of assembly tasks $j \in J = \{1, \ldots, n\}$ can be performed to simultaneously assemble v products $k \in K = \{1, \ldots, v\}$ of various types. A product completed on station i is transferred to the next station $i+1$ or another downstream station $l > i$, depending on the product assembly route, and it cannot revisit any station.

A feasible solution of the combined loading, routing and assembly plan selection problem must satisfy the following six basic types of constraints:

- Each assembly task must be assigned to at least one station.
- One assembly plan must be selected for each product.
- For each product and assembly plan selected all assembly tasks required must be completed.
- The total space required for the tasks assigned to each station must not exceed the station finite working space available.
- Tasks must be assigned to stations in such a way that for each assembly plan selected, precedence relations among the tasks are maintained in a unidirectional flow line with no revisiting of stations required.
- Each product must be successively routed to the stations where the required tasks have been assigned.

4.2.1 Problem formulation

In this section 0-1 integer programming formulation is presented for simultaneous loading, routing and assembly plan selection in a flexible assembly line. To model the problem the following decision variables are introduced (for notation used, see Table 4.1):

u_s = 1, if assembly plan $s \in S$ is selected (for the corresponding product k such that $s \in S_k$); otherwise $u_s = 0$

x_{ij} = 1, if task j is assigned to station $i \in I_j$; otherwise $x_{ij} = 0$

y_{iljs} = 1, if for assembly plan s, after completion of task j on station i, the product is transferred to station $l \geq i$ to perform the next task; otherwise $y_{iljs} = 0$

z_{ijs} = 1, if for assembly plan s task j is assigned to station i ; otherwise $z_{ijs} = 0$.

Let us notice that for each product $k \in K$, variables y_{iljs} (or z_{ijs}) must define a unique assembly route, since only one assembly plan $s \in S_k$ is selected for each $k \in K$ and subsets S_k are disjoint.

4.2 Loading, routing and assembly plan selection in a flexible assembly line

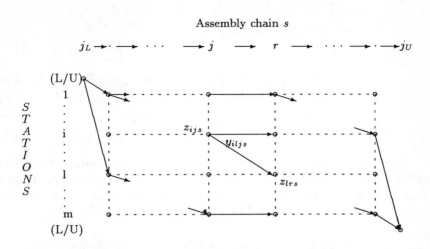

Fig. 4.5. Alternative assembly routes in a FAL

A bicriterion machine loading, product routing and assembly plan selection problem is formulated below as a bi-objective 0-1 integer program with a network flow structure embedded (see Fig. 4.5).

Model LRS': *Balancing station workloads and minimizing total transportation time in a flexible assembly line*

Minimize
$$P_{max}, Q_{sum} \tag{4.19}$$

subject to

$$\sum_{i \in I_j} \sum_{l \geq i} y_{iljs} = u_s; \quad s \in S, j \in T_s \tag{4.20}$$

$$\sum_{l \leq i} y_{lijs} - \sum_{l \geq i} y_{ilrs} = 0; \quad i \in I_r, (j,r) \in R_s, s \in S \tag{4.21}$$

$$\sum_{k \in K} \sum_{s \in S_k} \sum_{j \in J_k} \sum_{l \geq i} p_{jk} y_{iljs} \leq P_{max}; \quad i \in I \tag{4.22}$$

$$\sum_{i \in I} \sum_{l > i} \sum_{s \in S} \sum_{j \in T_s} q_{il} y_{iljs} = Q_{sum} \tag{4.23}$$

$$\sum_{i \in I_j} x_{ij} \geq 1; \quad j \in J \tag{4.24}$$

$$\sum_{j \in J} a_{ij} x_{ij} \leq b_i; \quad i \in I \tag{4.25}$$

$$y_{iljs} \leq x_{ij}; \quad i \in I_j, l \geq i, l \in I_r, (j,r) \in R_s, s \in S \tag{4.26}$$

100   4. Loading and Routing Decisions with Assembly Plan Selection

$$y_{iljs} \leq x_{lr}; \quad i \in I_j, l \geq i, l \in I_r, (j,r) \in R_s, s \in S \qquad (4.27)$$

$$y_{iljs} \leq u_s; \quad i \in I_j, l \geq i, l \in I_r, (j,r) \in R_s, s \in S \qquad (4.28)$$

$$\sum_{s \in S_k} u_s = 1; \quad k \in K \qquad (4.29)$$

$$y_{iljs} = 0; \quad i \in I, l < i, j \in T_s, s \in S \qquad (4.30)$$

$$u_s \in \{0,1\}; \quad \forall s \qquad (4.31)$$

$$x_{ij} \in \{0,1\}; \quad \forall i,j \qquad (4.32)$$

$$y_{iljs} \in \{0,1\}; \quad \forall i,l,j,s \qquad (4.33)$$

The constraints (4.20)–(4.29) play a similar role as constraints (4.2)–(4.11) in model **LRS**. In addition, however, they account for a unidirectional flow of products in a FAL. Furthermore, constraint (4.30) eliminates upstream flow of products in a unidirectional flow system.

4.2.2 A two-level loading, routing and assembly plan selection

An efficient solution to the bicriterion problem **LRS'** can be found by applying the two-level approach presented in Sect. 4.1.2, where first one solves the **LRS'** problem with the objective of minimizing P_{max} and with constraint (4.23) omitted.

Let us notice that the product assignment variables z_{ijs} and the assembly routing variables y_{iljs} are dependent via flow balance equations

$$z_{ijs} = \sum_{l \geq i} y_{iljs}; \quad j \in T_s, \ i \in I_j \qquad (4.34)$$

Using relation (4.34) the single objective problem of minimizing P_{max} can be reformulated into the following loading problem **L'**.

Model L': *Balancing station workloads in a flexible assembly line*

Minimize
$$P_{max} \qquad (4.35)$$

subject to

$$\sum_{i \in I_j} z_{ijs} = u_s; \quad s \in S, j \in T_s \qquad (4.36)$$

$$\sum_{k \in K} \sum_{s \in S_k} \sum_{j \in J_k} p_{jk} z_{ijs} \leq P_{max}; \quad i \in I \qquad (4.37)$$

$$z_{irs} \leq \sum_{l \leq i} z_{ljs}; \quad i \in I, s \in S, (j,r) \in R_s \qquad (4.38)$$

$$\sum_{j \in J} a_{ij} x_{ij} \leq b_i; \quad i \in I \qquad (4.39)$$

4.2 Loading, routing and assembly plan selection in a flexible assembly line

$$\sum_{i \in I_j} x_{ij} \geq 1; \qquad j \in J \qquad (4.40)$$

$$z_{ijs} \leq x_{ij}; \qquad s \in S, j \in T_s, i \in I_j \qquad (4.41)$$

$$x_{ij} = 0; \qquad j \in J, i \notin I_j \qquad (4.42)$$

$$z_{ijs} = 0; \qquad s \in S, j \notin T_s \text{ and/or } i \notin I_j \qquad (4.43)$$

$$\sum_{s \in S_k} u_s = 1; \qquad k \in K \qquad (4.44)$$

$$u_s \in \{0, 1\}; \qquad \forall s \qquad (4.45)$$

$$x_{ij} \in \{0, 1\}; \qquad j \in J, i \in I_j \qquad (4.46)$$

$$z_{ijs} \in \{0, 1\}; \qquad s \in S, j \in T_s, i \in I_j \qquad (4.47)$$

Constraint (4.38) for each assembly plan selected maintains the precedence relations among the tasks and ensures that a product does not revisit any station in a unidirectional flow system. If for assembly plan s, task r is assigned to station i, then task j that is to be performed immediately before r, i.e., $(j,r) \in R_s$, must be assigned to a station l such that $l \leq i$.

The loading problem **L'** can be solved by using the linear relaxation-based heuristic presented in Sect. 3.6.3, which finds a solution to the integer program **L'** starting from the optimal solution of the LP relaxation LP(**L'**) of **L'**.

In addition, the model of the loading problem **L'** can be further strengthened by the addition of valid constraints to the original formulation, which reduce the feasible region for the linear programming relaxation without eliminating the optimal integer solution. For example, the following equality implied by constraints (4.36) and (4.44) can be added

$$\sum_{s \in S_k} \sum_{i \in I_j} z_{ijs} = 1; \; k \in K, j \in J_k \qquad (4.48)$$

Let P_{max}^L be the solution value of P_{max} (4.35) obtained by applying the linear relaxation-based heuristic. Having solved problem **L'**, the bi-objective problem **LRS'** is next reduced into the following single objective simplified assembly routing and plan selection problem **RS'** of minimizing Q_{sum} for prefixed task assignments x_{ij}^L obtained for **L'** and with P_{max} bounded from above by P_{max}^L.

Model RS': *Minimizing total transportation time in a flexible assembly line for prefixed task assignments*

Minimize
$$Q_{sum} = \sum_{i \in I} \sum_{l > i} \sum_{s \in S} \sum_{j \in T_s} q_{il} y_{iljs} \qquad (4.49)$$

subject to (4.20, (4.21), (4.29), (4.30), (4.31), (4.33)

$$\sum_{k \in K} \sum_{s \in S_k} \sum_{j \in J_k} \sum_{l \geq i} p_{jk} y_{iljs} \leq P_{max}^L; \quad i \in I \qquad (4.50)$$

$$y_{iljs} \leq x_{ij}^L x_{lr}^L u_s; \quad i \in I_j, l \geq i, l \in I_r, (j,r) \in R_s, s \in S \qquad (4.51)$$

The objective Q_{sum} is a measure of the material handling system total workload. Constraint (4.50) defines an upper bound on each station workload. Constraint (4.51) is equivalent to variable upper bound constraints (4.26), (4.27) and (4.28) for prefixed task assignments x_{ij}^L.

The two-level approach for station loading, assembly routing and assembly plan selection problem in a FAL can be summarized by a figure similar to Fig. 4.3.

4.2.3 Numerical examples

In this subsection a simple numerical example is presented to illustrate the application of the 2-level approach for loading, routing and assembly sequence selection in a FAL.

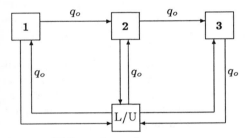

Fig. 4.6. FAL configuration

The FAL configuration for the example is provided in Fig. 4.6. The system is made up of $m = 3$ assembly stations $i = 1, 2, 3$ in series, and one L/U station directly connected with the assembly stations. The material handling system is unidirectional with $q_o = 2$ time units required for an AGV to move between any two neighbouring stations.

The production batch consists of $v = 5$ products of four types to be assembled of $n = 10$ types of components using $w = 10$ assembly plan where two alternative assembly sequences are available for each product. The available sequences $s \in S_k$ of tasks $j \in J_k$ required to assemble each product $k = 1, 2, 3, 4, 5$ are the same as in the example from Sect. 4.1.3 and so are the assembly times p_{jk}.

For each assembly task j the working space a_{ij} required for the corresponding component feeders is independent on the station i, i.e., $a_{ij} = a_j$, $\forall i \in I_j$, $j \in J$ ($a_{ij} = 0$ indicates that station i is incapable of performing task j):

4.2 Loading, routing and assembly plan selection in a flexible assembly line

$$[a_{ij}] = \begin{bmatrix} 1,2,3,1,2,3,0,0,0,0 \\ 0,0,0,1,2,3,1,2,3,5 \\ 1,2,3,0,0,0,1,2,3,5 \end{bmatrix}$$

The available working space for each station is: $b_1 = 10$, $b_2 = 10$, $b_3 = 10$.

First, an approximate solution of the loading problem **L'** strengthened by the addition of equality (4.48) has been determined applying the linear relaxation-based heuristic. A feasible solution has been found after 6 iterations of the heuristic. The violated valid inequalities (3.117) generated in each iteration and the final solution results obtained are presented below.

Iteration 1: $x_{1,1} + x_{1,2} + x_{1,3} + x_{1,5} + x_{1,6} \leq 4$
Iteration 2: $x_{3,7} + x_{3,8} + x_{3,9} + x_{3,10} \leq 3$
Iteration 3: $x_{3,1} + x_{3,8} + x_{3,9} + x_{3,10} \leq 3$
Iteration 4: $x_{2,4} + x_{2,5} + x_{2,7} + x_{2,8} + x_{2,10} \leq 4$, $x_{3,2} + x_{3,8} + x_{3,9} + x_{3,10} \leq 3$
Iteration 5: $x_{1,2} + x_{1,3} + x_{1,4} + x_{1,5} + x_{1,6} \leq 4$
Iteration 6: $x_{2,5} + x_{2,6} + x_{2,7} + x_{2,8} + x_{2,9} \leq 4$
Task assignments: $x_{1,1} = 1, x_{1,2} = 1, x_{1,3} = 1, x_{1,4} = 1, x_{2,4} = 1, x_{2,5} = 1, x_{2,6} = 1, x_{2,7} = 1, x_{2,8} = 1, x_{3,8} = 1, x_{3,9} = 1, x_{3,10} = 1$
Maximum workload: $P_{max}^L = 41$.

Next, given the above solution x_{ij}^L, P_{max}^L of the loading problem **L'**, the best assembly plans u_s and assembly routes y_{iljs} (and the corresponding product assignments z_{ijs} (4.34)) have been determined by solving the routing problem **RS'**. Direct application of an LP code to LP relaxation of **RS'** yields an integer solution with the maximum workload $P_{max} = 41$ and total transportation time $Q_{sum} = 38$. The assembly sequence and route selected for each product are shown in Fig. 4.7, where the number of the station selected for each task to be performed is additionally indicated in parentheses.

$k = 1, s = 1 : (L) \rightarrow 1(1) \rightarrow 2(1) \rightarrow 3(1) \rightarrow 4(1) \rightarrow 6(2) \rightarrow 8(2) \rightarrow (U)$
$k = 2, s = 3 : (L) \rightarrow 1(1) \rightarrow 2(1) \rightarrow 4(1) \rightarrow 5(2) \rightarrow 6(2) \rightarrow 7(2) \rightarrow 9(3) \rightarrow 10(3) \rightarrow (U)$
$k = 3, s = 5 : (L) \rightarrow 2(1) \rightarrow 3(1) \rightarrow 4(1) \rightarrow 5(2) \rightarrow 7(2) \rightarrow 8(3) \rightarrow 9(3) \rightarrow 10(3) \rightarrow (U)$
$k = 4, s = 7 : (L) \rightarrow 1(1) \rightarrow 3(1) \rightarrow 5(2) \rightarrow 6(2) \rightarrow 7(2) \rightarrow 8(3) \rightarrow 9(3) \rightarrow 10(3) \rightarrow (U)$
$k = 5, s = 9 : (L) \rightarrow 1(1) \rightarrow 3(1) \rightarrow 5(2) \rightarrow 6(2) \rightarrow 7(2) \rightarrow 8(3) \rightarrow 9(3) \rightarrow 10(3) \rightarrow (U)$

Fig. 4.7. Graph of assembly sequences and routes selected

Table 4.5 summarizes solution results for the example. It compares the results obtained using the 2-level approach and the linear relaxation heuristic with those determined by solving integer program **LRS'** with reduced single objective function $\lambda P_{max} + (1 - \lambda)Q_{sum}$ for $\lambda \in \{0, 0.5, 1\}$. The integer program **LRS'** was solved using discrete optimizer LINGO [115]. CPU time for the LINGO optimizer and the example was over one hour on a PC 486, whereas the computation time required for the 2-level approach with the linear relaxation heuristic was not greater than several seconds.

Table 4.5. Comparison of solution results

Objective function	2-level approach LP(**L'**), **RS'**	Integer program **LRS'** with single objective: $\lambda P_{max} + (1-\lambda)Q_{sum}$		
		$\lambda = 0$	$\lambda = 0.5$	$\lambda = 1$
P_{max}	41	52	39	39
Q_{sum}	38	38	38	40
Selected sequences $s =$	1,3,5,7,9	1,3,5,7,9	2,4,5,7,9	1,3,5,7,9

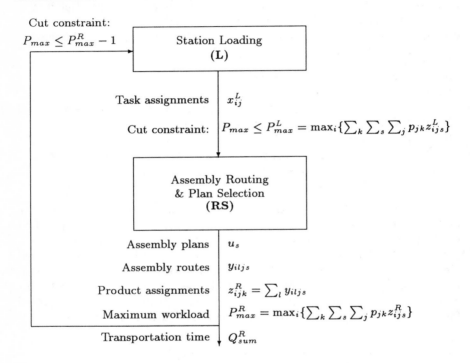

Fig. 4.8. Hierarchical task assignment, assembly routing and plan selection

The two-level approach proposed for the bi-objective loading, routing and assembly plan selection enables a network flow structure of the routing and plan selection subproblem to be efficiently exploited. However, it is a pure top down approach and the solution obtained at the lower level has no effect on the upper level problem. To avoid that disadvantage, an improved hierarchical approach shown in Fig. 4.8 can be considered as an alternative optimization scheme. In the hierarchical framework proposed in Fig. 4.8 the solution value obtained at one level generates a cut constraint for the other level. The cuts are based on the current upper bounds on the value of each objective function, which may further improve performance of the lexicographic approach proposed for the multiobjective optimization. The new framework can be in-

4.2 Loading, routing and assembly plan selection in a flexible assembly line

corporated into an interactive optimization scheme where the most preferred final solution can be selected based on the decision maker's preferences.

5. Loading and Scheduling in Flexible Assembly Cells

In this chapter mathematical programming formulations are presented for simultaneous loading and scheduling in flexible assembly cells. The formulations are illustrated with practical applications in mechanical part assembly in a robot assembly cell and in printed circuit board assembly on a component placement machine. The models proposed enable an allocation of production tasks and resources to be made along with a detailed sequencing and scheduling decisions, and by this lead to a global optimization of the assembly process.

5.1 Loading and scheduling in a robot assembly cell

This section presents a monolithic approach developed in [83] for simultaneous loading and scheduling in a robot assembly cell. The monolithic approach leads to well-defined model formulation which enables an overall system optimization to be achieved, e.g., [43, 50]. The loading and scheduling problem is formulated as a mixed 0-1 linear program which can be solved using commercially available optimization software. The monolithic approach is essentially different from a widely used hierarchical approach, e.g., [68]. The hierarchical approach partitions the loading and scheduling problem into a hierarchy of subproblems to be solved sequentially (see Fig. 2.13) and at best can only optimize each of these subproblems. In contrast to monolithic model, the hierarchical approach results in suboptimization of the loading and scheduling decisions. The main disadvantage of the monolithic approach is the necessity to solve large scale integer programming models. However, recent theoretical advances in integer programming and continuous progress observed in computer technology have resulted in advanced commercial software for discrete optimization that can handle large scale mathematical programs in a reasonable computation time. Real-time applications of monolithic approaches based on integer programming formulations are envisaged in production planning and scheduling in the nearest future.

Let us consider an assemby cell in which a set of operations is performed using a set of resources to assemble final products.

A resource (facility) $i \in I$ may be an assembly robot, a buffer, a component feeder, or any major subspace in a robot working space. The latter

5. Loading and Scheduling in Flexible Assembly Cells

resource can be defined in a multirobot assembly cell to avoid robot collision. The subspace can be used by at most one robot at a time and hence collision of robots is avoided, see Fig. 1.9.

An operation $j \in J$ is any physical activity or group of activities that is carried out by utilizing at least one resource.

Each operation can be performed in many ways, called performing modes. Performing modes $k \in K$ are determined by the resource requirement. A resource requirement includes the set of resources that are needed, and the timings of the resources utilization. Each resource may be used during a single time segment, which might be any portion of the total processing time of operation under the performing mode selected.

A pair (j, k) of an operation j with selected performing mode k (and its resource requirements) forms a set of tasks that every resource required is assigned to perform. As a result each resource i will be assigned a sequence of tasks that are indexed by $l \in L_i$, where L_i is the set of such tasks.

The notation used for the model formulation is presented in Table 5.1.

Table 5.1. Notation

		Indices
i	=	resource, $i \in I$
j	=	operation, $j \in J$
k	=	performing mode (K_j = the set of performing modes of operation j)
l	=	task (L_i = the set of tasks that can be performed using resource i)
		Input parameters
d_{i,j_1,k_1,j_2,k_2}	=	the minimal delay required by resource i to perform (j_2, k_2) after completing (j_1, k_1)
p_{jk}	=	processing time of operation j under mode k
$p_{ijk}^{(s)}$	=	start time of using resource i when operation j is performed by mode k, after this (j, k) has begun
$p_{ijk}^{(f)}$	=	finish time of using resource i when operation j is performed by mode k, after this (j, k) has begun
I_{jk}	=	the subset of resources that are required to perform operation j by mode k
J_i	=	the subset of (j, k)s that require the resource i
D_i	=	$\{[(j_1, k_1), (j_2, k_2)] : (j_1, k_1) \in J_i, (j_2, k_2) \in J_i, d_{i,j_1,k_1,j_2,k_2} > 0\}$ – the set of pairs of operations $[(j_1, k_1), (j_2, k_2)]$ such that use the same resource i and if selected as consecutive tasks of i will necessitate its delay by d_{i,j_1,k_1,j_2,k_2} time units
R	=	the set of immediate predecessor-successor pairs of operations
		Decision variables
s_j	=	start time of operation j
s_{il}	=	start time of performing task l by resource i
x_{ijkl}	=	1, if operation j performed by mode k is assigned as the lth task of resource i; otherwise $x_{ijkl} = 0$
z_{il}	=	the minimal delay time between completion of task $l - 1$ and beginning of performing of task l by resource i

5.1 Loading and scheduling in a robot assembly cell

As an illustrative example consider an operation for which three types of resources are required: an assembly robot, a part feeder and a space at the part feeder. The operation consists of moving a base part from an input buffer to a temporary storage area at the part feeder, positioning the base part, gripping the required component from the feeder and assemble it with the base part. The operation is carried by the robot ($i = 1$) and in addition, requires the temporary stoarge area ($i = 2$) and the part feeder ($i = 3$) to be used during some portions of the total processing time, see Fig. 5.1.

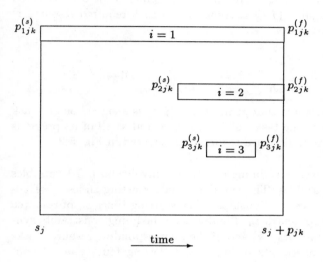

Fig. 5.1. Resource requirements for operation j under mode k

5.1.1 Loading and scheduling constraints

Loading constraints. The loading constraints guarantee a feasible selection of one performing mode for each operation and the assignment of pairs (j, k) = (operation, performing mode) as the consective tasks for the resources required by each (j, k), see [83]

- Mode selection constraints

$$\sum_{k \in K_j} \sum_{l \in L_{i'}} x_{i'jkl} = 1; \ \forall j \in J \tag{5.1}$$

Constraint (5.1) allows only one performing mode for each operation to be selected. If for operation j mode k is selected, then

$$\sum_{l \in L_i} x_{ijkl} = 1; \; \forall i \in I_{jk}.$$

It is enough to use only one resource $i' \in I_{jk}$ to detect the selection of mode k for operation j.

– Resource allocation constraints

$$\sum_{l \in L_{i'}} x_{i'jkl} - \sum_{l \in L_i} x_{ijkl} = 0; \; \forall i \in I_{jk} \setminus \{i'\}, \; k \in K_j, \; j \in J \qquad (5.2)$$

Constraint (5.2) assures that all resource requirements for each selected (j, k) are met by assigning (j, k) as some task of each required resource i.

– Task assignment constraints

$$1 \geq \sum_{(j,k) \in J_i} x_{ijk1} \geq \sum_{(j,k) \in J_i} x_{ijk2} \geq \cdots \geq \sum_{(j,k) \in J_i} x_{ijk|L_i|}; \; \forall i \in I \qquad (5.3)$$

Constraints (5.3) guarantee that at most one (j, k) is assigned as one task l of resource i and that the task l of i is not used unless all of its previous task indices are utilized. The idea is further illustrated in Fig. 5.2.

Scheduling constraints. Scheduling constraints involve both 0-1 variables and continuous time variables. The variables relate starting times s_j of operations to their precedence relationships. Also, starting times s_{il} of resource tasks are related to setup delays z_{il}. Finally, the scheduling constraints synchronize the starting times of (j, k) and of their corresponding resource tasks (i, l). The scheduling constraints include the following four types of constraints, see [83].

– Precedence constraints

$$s_{j_1} + \sum_{k \in K_j} p_{j_1 k} \sum_{l \in L_{i'}} x_{i'j_1 kl} \leq s_{j_2}; \; \forall (j_1, j_2) \in R \qquad (5.4)$$

Constraint (5.4) maintains the precedence relationships among all operations. For each pair $(j_1, j_2) \in R$ of consecutive operations, their corresponding starting times differ of at least the duration of operation j_1.

– Changeover time constraints

$$z_{il} \geq d_{i,j_1,k_1,j_2,k_2}(x_{ij_1 k_1 l-1} + x_{ij_2 k_2 l} - 1); \qquad (5.5)$$
$$\forall [(j_1, k_1), (j_2, k_2)] \in D_i, \; l = 2, \ldots, |L_i|, \; i \in I$$

Constraint (5.5) introduces the minimal changeover time delay required between tasks $(l-1)$ and l of resource i. It allows for consideration of any changeover, setup or safety time gaps, if required.

5.1 Loading and scheduling in a robot assembly cell

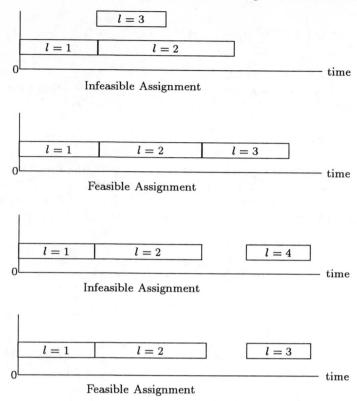

Fig. 5.2. Examples of task assignment [83]

- Task sequencing constraints

$$s_{il} \geq s_{il-1} + z_{il} + \sum_{(j,k)\in J_i} (p_{ijk}^{(f)} - p_{ijk}^{(s)})x_{ijkl-1}; \quad (5.6)$$
$$\forall l = 2, \ldots, |L_i|, \ i \in I$$

Constraint (5.6) guarantees that the time difference between starting times s_{il} and s_{il-1} is at least the sum of the processing time required for task $(l-1)$ and the minimal delay z_{il}.
- Task and operation starting times interdependence

$$Q\left(\sum_{k\in K_j} x_{ijkl} - 1\right) \leq s_j - s_{il} + p_{ijk}^{(s)} \sum_{k\in K_j} x_{ijkl}$$

$$\leq Q\left(1 - \sum_{k \in K_j} x_{ijkl}\right) ; \ \forall i \in I, \ j \in J, \ l \in L_i \qquad (5.7)$$

where Q is a sufficiently large number.
Constraint (5.7) ties starting time s_j of operation j with starting time s_{il} of the lth task of resource i, if (j, k) was assigned to i as its lth task.

5.1.2 Objective functions

This subsection presents four typical objective functions for the simultaneous loading and scheduling in a robot assembly cell. Constraints (5.1) – (5.7) are required for any of the objective function selected. Additional constraints needed for any particular objective are presented along with that objective.

– *Minimization of schedule length C_{max}*

$$C_{max} = \max_j \{s_j + \sum_{k \in K_j} p_{jk} \sum_{l \in L_{i'}} x_{i'jkl}\} \qquad (5.8)$$

The scheduling problem can be formulated as follows:

Minimize
$$C_{max} \qquad (5.9)$$

subject to (5.1) – (5.7) and

$$s_j + \sum_{k \in K_j} p_{jk} \sum_{l \in L_{i'}} x_{i'jkl} \leq C_{max}; \qquad \forall j \qquad (5.10)$$

$$s_j \geq 0; \qquad \forall j \qquad (5.11)$$

$$s_{il} \geq 0; \qquad \forall i, l \qquad (5.12)$$

$$x_{ijkl} \in \{0, 1\}; \qquad \forall i, j, k, l \qquad (5.13)$$

$$z_{il} \geq 0; \qquad \forall i, l \qquad (5.14)$$

– *Maximization of the system performance*
The goal is to maximize the sum of evaluations w_{jk} of each operation j with its chosen performance mode k

$$\sum_{j \in J} \sum_{k \in K_j} w_{jk} \sum_{l \in L_{i'}} x_{i'jkl} \qquad (5.15)$$

where w_{jk} is the weight assigned to operation j performed by mode k.

– *Minimization of resource costs*
  The total cost of using the resources in the system consists of fixed cost (e.g. installation cost, depreciation, and overhead costs) and variable cost. The fixed cost of a resource is incurred if at least one (the first) task is assigned to that resource. Variable costs are measured on the basis of the net processing time during which a resource is used. For example the objective function can be expressed as below.
  Minimize

$$\sum_{i \in I} \left(c_i \sum_{(j,k) \in J_i} x_{ijk1} + g_i \sum_{(j,k) \in J_i} \sum_{l \in L_i} (p_{ijk}^{(f)} - p_{ijk}^{(s)}) x_{ijkl} \right) \quad (5.16)$$

  where: c_i – fixed cost of resource i, g_i – variable cost of resource i per unit of time.

– *Workload balancing*
  A workload balancing objective aims at allocating similar working loads to a subset of resources of the same type (e.g. assembly robots). For example the workload balancing problem can be formulated as below.
  Minimize
$$W_2 - W_1 \quad (5.17)$$
  subject to (5.1) – (5.7), (5.11) – (5.14) and

$$W_1 \leq \sum_{(j,k) \in J_i} \sum_{l \in L_i} (p_{ijk}^{(f)} - p_{ijk}^{(s)}) x_{ijkl} \leq W_2; \quad \forall i \in E \subset I \quad (5.18)$$

$$W_1 \geq 0, \, W_2 \geq 0 \quad (5.19)$$

  where: W_1 and W_2 are two additional decision variables representing the lower and the upper bounds on workload, and E is the subset of resources of the same type.

In [83] the monolithic model presented above is applied for practical loading and scheduling problem of a multirobot assembly cell, namely an Ignitor Assembly Cell with two robots operating simultaneously, described in Sect. 1.3.1. The problem was solved using LINDO discrete optimizer and the optimal assembly schedules were determined for the four objective functions described above. The results have indicated that monolithic mixed-integer programming formulations can be used in practice for simultaneous loading and scheduling of robot assembly cells.

5.2 Loading and sequencing in printed circuit board assembly

In this section the machine loading and assembly sequencing problem is presented for the PCB assembly. Let us consider again the Fuji CP II component placement machine described in Sect. 1.3.2 (Fig. 1.12). The PCB assembly process can be decomposed into the following three subproblems (cf., Sect. 2.5.3).

1. Table tour – the placement sequence is specified to determine the movement of the table.
2. Component assignment – the assignment of component types to magazine slots is determined.
3. Component retrieval sequence – the movement of the magazine rack is determined if more than one slot is assgned the same component type.

The placement sequence problem can be formulated as a Travelling Salesman Problem (TSP) where each location on the board must be visited once, and where the distance between two locations is a function of the time required to move the table between them. The problem objective is to minimize total travel time, see [20]. For the given placement sequence, next the component assignment problem and the component retrieval sequence problem is solved, collectively known as the *component retrieval problem*.

5.2.1 Component retrieval problem

Table 5.2. Notation

	Indices	
i	=	placement location on the PCB, $i \in I$
j, k	=	magazine positions, $j, k \in J$
p	=	component type, $p \in P$
	Input parameters	
a_{ip}	=	1, if placement location i on the PCB requires component type p; otherwise $a_{ip} = 0$
I	=	ordered set of locations (X, Y) on the PCB, $n_I = |I|$
I_p	=	set of placement locations (X, Y) on the PCB that require component type p
J	=	set of magazine positions, $n_J = |J|$
P	=	set of component types, $n_P = |P|$
n_D	=	number of placement head positions between the placement station and the gripping station (for the Fuji CP II, $n_D = 6$)
$t_{jk}(i-1, i)$	=	incremental amount of time required to complete the retrieval of the ith component from position k on the magazine rack given that the $i-1$st component was just retrieved from magazine position j
η_p	=	maximum number of magazine positions component type p may occupy
	Decision variables	
x_{ij}	=	1, if the component type associated with the ith placement is in magazine position j; otherwise $x_{ij} = 0$
y_{jp}	=	1, if component type p is in magazine position j; otherwise $y_{jp} = 0$

5.2 Loading and sequencing in printed circuit board assembly

Let us note that the ordered set I of placement locations (X, Y) on the board has been obtained by earlier solving the corresponding TSP problem of the component placement sequence which determines the movement of the table.

This subsection presents mathematical formulation of the component retrieval problem developed in [20]. The notation used to formulate the problem is shown in Table 5.2.

The problem has been formulated under the following assumptions:

- for the PCBs under consideration, all tape reels are of equal size,
- at the beginning the placement head carousel is empty and the magazine rack and table are in the proper positions for retrieval and placing the first componenet,
- the machine cannot begin assembly of the next board until all operations on the current board are completed,
- ejection rates during inspections are negligible.

The component retrieval problem is formulated below as a quadratic integer program ([20]).

Minimize

$$f(x,y) = \sum_{i=2}^{n_I}\sum_{j=1}^{n_J}\sum_{k=1}^{n_J} t_{jk}(i-1,i)x_{i-1\,j}x_{ik} + \tau \qquad (5.20)$$

subject to

$$\sum_{j=1}^{n_J} x_{ij} = 1; \qquad i \in I \qquad (5.21)$$

$$\sum_{i=1}^{n_I} a_{ip}x_{ij} \leq |I_p|y_{jp}; \qquad j \in J,\ p \in P \qquad (5.22)$$

$$\sum_{j=1}^{n_J} y_{jp} \leq \eta_p; \qquad p \in P \qquad (5.23)$$

$$\sum_{p=1}^{n_P} y_{jp} \leq 1; \qquad j \in J \qquad (5.24)$$

$$x_{ij} \in \{0,1\}; \qquad i \in I,\ j \in J \qquad (5.25)$$

$$y_{jp} \in \{0,1\}; \qquad j \in J,\ p \in P \qquad (5.26)$$

The objective function (5.20) represents the total time required to grip and place all the component on the board. For a given component placement sequence I, the total time is find by summing the time between consecutive component retrievals, and then adding some fixed time τ for placing the final n_D components.

Constraint (5.21) ensures that each location on the PCB receives a component from exactly one magazine position. Similarly, (5.22) guarantees that the components are retrieved from the magazine positions that have been assigned the required component type. Constraint (5.23) limits the number of magazine positions to which a component type may be assigned. Constraint (5.24) guarantees that each position on the magazine rack is allocated no more than one component type.

Problem (5.20) – (5.26) is a quadratic integer program which is difficult to solve with standard techniques even for small size problems. In addition, the term $t_{jk}(i-1,i)$ in the objective function (5.20) is dependent upon the component retrieval sequence x and the magazine assignments y, yet to be determined. In order to determine the values of $t_{jk}(i-1,i)$, all feasible combinations of x and y should be examined.

In [20] an alternative approach is proposed based on the decomposition of problem (5.20) – (5.26) which has been achieved by relaxing the complicating constraint (5.22) using a Lagrangian relaxation scheme.

6. Production Scheduling in Flexible Assembly Lines

This chapter is devoted to production scheduling in flexible assembly lines. A *flexible assembly line* (or a mixed-model assembly line) is a unidirectional flow type production system. It consists of several assembly stages in series, either separated by finite interstage buffers or with no intermediate buffers between the stages. Each stage has one or more identical parallel machines. The line can produce several different part types simultaneously. Each part must be assembled by at most one machine in each stage, but some parts may skip some stages. In the line with limited intermediate buffers limited queueing between the assembly stages may occur, whereas in the line with no buffers queueing is not allowed.

A flexible assembly line represents a combination of traditional flowshop in which there is only one machine in each stage, and the identical parallel machine shop in which there is only one processing stage with more than one parallel machines. However, unlike in the flexible assembly line, the traditional flowshop requires every part to visit every machine and unlimited queueing between the machines is allowed, e.g., [27]. The flexible assembly line with unlimited intermediate buffers has also been termed a flexible flow shop or a hybrid flow shop, e.g., [44].

The research on the development of scheduling algorithms for flexible assembly lines is mostly restricted to heuristics which hopefully obtain good solutions in reasonable computing times, e.g., [118, 129]. For example, in [129] a heuristic scheduling algorithm is proposed primarily to minimize the makespan and secondarily to minimize the in-process inventory. The approach of the algorithm is based on the decomposition of the problem into three subproblems to be solved sequentially: (i) machine allocation (determining which parts will visit each individual machine at each stage); (ii) part input sequencing (specifying the order in which the parts should enter the line); (iii) timing (specifying the times at which the parts should enter the line as well as all start and completion times for all parts in every stage).

The scheduling algorithms presented in this chapter aim at minimizing the makespan while the in-process inventory is limited by the intermediate buffer capacities, if there are buffers in the line. The algorithms proposed are part-by-part heuristics, in which during every iteration a complete assembly schedule is determined for one part. The selection of the part and its complete schedule are based on the cumulative partial schedule obtained for all parts selected so far. The decisions in every iteration are made using a local optimization procedure aimed at minimizing total idle time along the route of the selected part. For this reason, the heuristics will be called *Route Idle Time Minimization* (RITM) [98] or *Route Idle Time Minimization - No Store* (RITM-NS) [101, 104], respectively for the line with limited intermediate buffers or the line with no in-process buffers.

The RITM and RITM-NS heuristics use a push-type scheduling strategy. An opposite pull-type strategy is applied for the Just-In-Time and multilevel scheduling of flexible assembly lines presented in Sect. 6.4 and 6.5.

6.1 Flexible assembly line with limited intermediate buffers

In this section the problem of scheduling a flexible assembly line with limited intermediate buffers is described and its mathematical formulation is presented.
Notation used to formulate the scheduling problem and the algorithm is introduced in Table 6.1.

The flexible assembly line study consists of $S \geq 2$ assembly stages in series with limited in-process buffers between the successive stages. Each stage i ($i = 1, \ldots, S$) is made up of $M_i \geq 1$ identical parallel machines. Ahead of each stage i ($i = 2, \ldots, S$) there are B_i buffers, where each buffer can hold one part at a time (see Fig. 6.1).

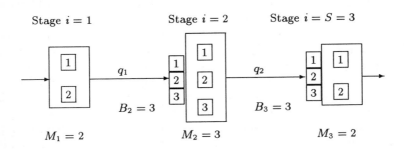

Fig. 6.1. Example of a three-stage flexible assembly line

The system produces N different part types. Let $p_{ij} \geq 0$ be the assembly time in stage i of part type j ($j = 1, \ldots, N$) and let q_i be the transportation time required to transfer a part from stage i to stage $i + 1$.

A part completed in stage i is transferred either directly to an available machine in the next stage $i + 1$ (or another downstream stage depending on

6.1 Flexible assembly line with limited intermediate buffers

Table 6.1. Notation

		Indices
i	=	assembly stage, $i = 1, \ldots, S$
j	=	part type, $j = 1, \ldots, N$
k	=	kth position in the loading sequence, $k = 1, \ldots, P$
		Input parameters
d_j	=	demand for part type j ($j = 1, \ldots, N$)
B_i	=	number of buffers ahead of stage i ($i = 2, \ldots, S$)
M_i	=	number of parallel machines in stage i ($i = 1, \ldots, S$)
P	=	total number of parts in the production program, $P = \sum_{j=1}^{N} d_j$
p_{ij}	=	assembly time of part type j in stage i
q_i	=	transportation time from stage i to stage $i+1$
w_i	=	average machine workload in stage i ($i = 1, \ldots, S$), $w_i = \sum_{j=1}^{N} p_{ij} d_j / M_i$
w^*	=	average machine workload in the bottleneck stage i^*, $w^* = w_{i^*} = \max_{1 \leq i \leq S} w_i$
		Decision variables
j_k	=	part number k ($k = 1, \ldots, P$) in the loading sequence $[j_1, j_2, \ldots, j_P]$
s_{ik}	=	start time of part j_k on machine in stage i ($i = 1, \ldots, S$)
c_{ik}	=	completion time of part j_k on machine in stage i ($i = 1, \ldots, S$)
r_{ik}	=	release time of part j_k from stage i ($i = 1, \ldots, S$)
		Auxiliary variables
t_{ik}	=	idle time on machine in stage i incurred by part j_k (waiting time for start of assembly part j_k and duration of machine blocking time by finished part j_k)
X_{bik}	=	total elapsed time, when buffer b ($b = 1, \ldots, B_i$) ahead of stage i is available after the first k parts have been scheduled
x_{ik}	=	the earliest time at which a buffer ahead of stage i is available after the first k parts have been scheduled, $x_{ik} = \min_{1 \leq b \leq B_i} \{X_{bik}\}$
$b(i,k)$	=	buffer ahead of stage i with the earliest available time after the first $k-1$ parts have been scheduled, $b(i,k) = \arg\min_{1 \leq b \leq B_i} \{X_{bik-1}\}$
Y_{mik}	=	total elapsed time, when machine m ($m = 1, \ldots, M_i$) at stage i is available for assignment after the first k parts have been scheduled
y_{ik}	=	the earliest time at which a machine in stage i is available for assignment after the first k parts have been scheduled, $y_{ik} = \min_{1 \leq m \leq M_i} \{Y_{mik}\}$
$m(i,k)$	=	machine in stage i with the earliest available time after the first $k-1$ parts have been scheduled, $m(i,k) = \arg\min_{1 \leq m \leq M_i} \{Y_{mik-1}\}$

the part assembly route) or to a buffer ahead of that stage. If all B_{i+1} buffers ahead of stage $i+1$ are occupied, then the machines in stage i are blocked by the completed parts until a buffer is available.

Given is a production program $(d_1, \ldots, d_N)$, where d_j ($j = 1, \ldots, N$) is the required number of parts of type j. The problem objective is to determine an assignment of all $P = \sum_{j=1}^{N} d_j$ parts to machines in each stage over a scheduling horizon in such a way as to complete the production program in a minimum time, that is, to minimize the makespan $C_{max} = \max_{1 \leq i \leq S}(c_{iP})$, where c_{iP} denotes the completion time in stage i of the last Pth part.

Production schedule for all P parts is specified by the loading sequence $[j_1, j_2, \ldots, j_P]$ in which the parts enter the line as well as the times required for detailed scheduling of each individual part. In particular, for each part j_k entering the line at position k ($k = 1, \ldots, P$) its start time s_{ik}, completion time c_{ik}, and release time r_{ik} in each stage i should be determined. These variables satisfy the following formulas:

$$s_{ik} = \max\{r_{i-1,k} + q_{i-1};\ y_{i,k-1}\};\ i = 2, \ldots, S \quad (6.1)$$

$$s_{1k} = y_{1,k-1}$$

$$c_{ik} = s_{ik} + p_{ij_k};\ i = 1, \ldots, S \quad (6.2)$$

$$r_{ik} = \max\{c_{ik},\ x_{i+1,k-1} - q_i\};\ i = 1, \ldots, S-1 \quad (6.3)$$

$$r_{Sk} = c_{Sk}$$

$$x_{ik} = \min_{1 \leq b \leq B_i}\{X_{bik}\};\ i = 2, \ldots, S \quad (6.4)$$

$$y_{ik} = \min_{1 \leq m \leq M_i}\{Y_{mik}\};\ i = 1, \ldots, S \quad (6.5)$$

where the earliest available times X_{bik} for buffers and Y_{mik} for machines are calculated iteratively as follows

$$X_{bik} = \begin{cases} X_{bik-1} & if\ b \neq b(i,k) \\ s_{ik} & if\ b = b(i,k) \end{cases} \quad (6.6)$$

$$Y_{mik} = \begin{cases} Y_{mik-1} & if\ m \neq m(i,k) \\ r_{ik} & if\ m = m(i,k) \end{cases} \quad (6.7)$$

The formulas (6.6) and (6.7) were derived under the assumption that in every stage i part j_k is assigned to the buffer $b(i,k)$ and the machine $m(i,k)$ with the earliest available times (for additional interpretation of the variables introduced, see Fig. 6.2).

Finally, notice that idle time t_{ik} incurred in stage i by part j_k consists of two components:
(i) time $(s_{ik} - y_{i,k-1})$ of waiting for start of assembly part j_k and
(ii) time $(r_{ik} - c_{ik})$ of machine blocking by finished part j_k.
Hence, t_{ik} can be expressed as follows

$$t_{ik} = (s_{ik} - y_{i,k-1}) + (r_{ik} - c_{ik}) = r_{ik} - y_{i,k-1} - p_{ij_k};\ i = 1, \ldots, S \quad (6.8)$$

6.1 Flexible assembly line with limited intermediate buffers

In order to minimize the makespan C_{max}, the RITM heuristic presented in the next section assigns parts to the earliest available machines and aims at minimizing the total idle time $\sum_{i=1}^{S}\sum_{k=1}^{P} t_{ik}$ along the routes of all P parts.

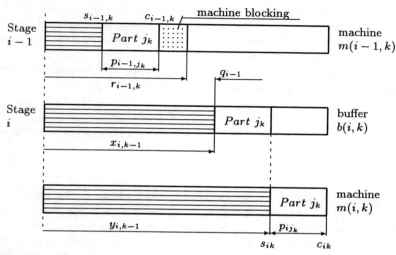

Fig. 6.2. A partial schedule for part j_k

6.1.1 Scheduling algorithm

The algorithm presented in this section is a part-by-part heuristic in which during every iteration a complete assembly schedule is determined for one part. The selection of the part and its complete schedule are based on the cumulative partial schedule obtained for all parts selected so far. The decisions in every iteration are made using a local optimization procedure aimed at minimizing total idle time along the route of the selected part. Therefore, the heuristic is called *Route Idle Time Minimization* (RITM).

First, let us recall that the algorithm proposed in [51] for two and three stage flowshops leads to the following observation, e.g., [118, 129]:

A "good" input sequence for a multistage flowshop will begin with parts with greater processing times in the final stages, continue with parts with processing times about the same in all the stages, and finish with parts with greater processing times in the beginning stages.

In order to take into account the impact of the distribution of assembly times among the successive stages on the actual flow time of a part, let us introduce the total weighted assembly time $\bar{p}_j$ for each part type j ($j = 1, \ldots, N$)

122   6. Production Scheduling in Flexible Assembly Lines

$$\overline{p}_j = \sum_{i=1}^{S}[1 + a_1(i-1)]p_{ij} \qquad (6.9)$$

where $a_1 \in [0,1]$ is a constant.
For each part type j the further toward the end of the line is stage i, the greater is the weight assigned to the assembly time p_{ij}. In this way parts with greater assembly times in the downstream stages are considered as if they required greater total assembly time. Such parts should enter the line at the beginning of the loading sequence.
In the algorithm RITM presented below all part types are initially ordered according to nonincreasing total weighted assembly times $\overline{p}_j$, that is,

$$\overline{p}_1 \geq \overline{p}_2 \geq \cdots \geq \overline{p}_N \qquad (6.10)$$

In the sequel a part type j (with the jth largest time $\overline{p}_j$) is assigned a weight

$$\varrho_j = a_2^{N-j} \qquad (6.11)$$

where $a_2 \geq 1$ is a constant.

The algorithm RITM is a single-pass part-by-part heuristic in which the loading sequence and the corresponding complete schedule are determined once. During every iteration a part for loading into the system is chosen as well as its complete assembly schedule is determined. The decisions in every iteration are made based on the complete assembly schedule determined for each part type waiting for loading.
Given a cumulative partial schedule for $(k-1)$ parts selected so far, first the best route along the line is found as a sequence

$$[m(1,k), b(2,k), m(2,k), \ldots, b(S,k), m(S,k)] \qquad (6.12)$$

of S machines and $S-1$ buffers (one in every stage) with the earliest available times. For each part type waiting for entering the line the complete assembly schedule is determined along the best route obtained. To evaluate the assembly schedule for each part type considered for loading, the total duration of weighted idle time $\overline{t}_j$ along the route is determined

$$\overline{t}_j = \sum_{i=1}^{S} t_{ij}/\varrho_j \qquad (6.13)$$

In equation (6.13) the total idle time for each part type j is divided by its weight ϱ_j. In this way total idle time incurred by part type j with a higher weight ϱ_j, that is, with a higher total weighted assembly time $\overline{p}_j$ is reduced. As a result such a part is given an additional priority to enter the line.
Finally, the part with the smallest total weighted idle time $\overline{t}_j$ is selected for loading and its complete assembly schedule is added to the cumulative partial schedule obtained so far.
Description of the algorithm RITM is given below (J denotes the set of part types loaded in the required number of parts).

Algorithm RITM

STEP 0. Determining weights for part types
1. Choose values of constant parameters a_1, a_2.
2. Order all part types according to nonincreasing values of total weighted assembly times $\bar{p}_j$, (6.10).
   Assign weight $\varrho_j = a_2^{N-j}$ to part type j (with the jth largest value of $\bar{p}_j$).
3. Set:
   $X_{bi0} = 0$, $b = 1, \ldots, B_i$, $i = 2, \ldots, S$,
   $Y_{mi0} = 0$, $m = 1, \ldots, M_i$, $i = 1, \ldots, S$,
   $b(i,1) = 1$, $i = 2, \ldots, S$, $m(i,1) = 1$, $i = 1, \ldots, S$,
   $J = \emptyset$, $k = 1$.

STEP 1. Selection of a part type for loading
1. For each part type $j \notin J$ waiting for loading determine start time s_{ij}, (6.1), completion time c_{ij}, (6.2), release time r_{ij}, (6.3), and duration t_{ij}, (6.8) of machine idle time, in every stage i on machine $m(i,k)$ with the earliest available time y_{ik-1}.
   Next, determine total weighted idle time $\bar{t}_j$ (6.13).
2. Select for loading such a part j_k that minimizes the total weighted idle time, that is
   $$j_k = \arg\min_{j \notin J}\{\bar{t}_j\}$$

STEP 2. Determining complete assembly schedule for the selected part
For the selected part j_k determine complete assembly schedule (times s_{ik}, (6.1) c_{ik}, (6.2) r_{ik}, (6.3)) by assigning it in every stage i to buffer $b(i,k)$ and machine $m(i,k)$ with the earliest available times, respectively x_{ik-1},(6.4), and y_{ik-1},(6.5), after the first $k-1$ parts have been scheduled. The assembly schedule for part j_k add to the cumulative partial schedule for the first $(k-1)$ parts.

STEP 3. Checking the state of completion the production program
For $j = j_k$ set $d_j = d_j - 1$. If $d_j = 0$, then set $J = J \bigcup \{j\}$.
If $J = \{1, \ldots, N\}$, then terminate.
Otherwise determine for each buffer b ($b = 1, \ldots, B_i$, $i = 2, \ldots, S$) and for each machine m ($m = 1, \ldots, M_i$, $i = 1, \ldots, S$) the earliest available time, respectively X_{bik},(6.6) and Y_{mik}, (6.7) after the first k parts have been scheduled.
For each stage i find buffer $b(i, k+1) = \arg\min_{1 \leq b \leq B_i}(X_{bik})$ and machine $m(i, k+1) = \arg\min_{1 \leq m \leq M_i}(Y_{mik})$ with the earliest available time x_{ik}, (6.4), y_{ik}, (6.5), respectively.
Set $k = k + 1$ and go to *STEP 1*.

To determine the computational complexity of the algorithm RITM notice that the algorithm requires P iterations and in every iteration a complete assembly schedule is computed for one part according to the following procedure.

First, the best route is found as a sequence of at most S machines and $S-1$ buffers in the successive stages with the earliest available times. This step requires $O(M+B)$ computations since at most all M machines and B buffers are considered ($M = \sum_{i=1}^{S} M_i$ and $B = \sum_{i=2}^{S} B_i$ is the total number of machines and buffers, respectively).

Then, for each of the remaining at most N part types, the complete assembly schedule is computed based on the best route found. This step requires $O(NS)$ computations to determine the times for the N flowshop schedules on S machines. The best assembly schedule (and by this a part for loading) with the smallest total weighted idle time is next chosen and added to the cumulative partial schedule for all parts loaded so far.

Since the above procedure which requires $O(M+B+NS)$ computations is repeated in each of the P iterations, and the number of part types $N \leq P$, the computational complexity of the algorithm RITM is $O(P^2 S)$.

In order to evaluate the effectiveness of the proposed heuristic algorithm, one would need to find the value of the optimal makespan. However, no exact optimization algorithm for this problem is available. Therefore, the following lower bound on makespan, based on average machine workload w^* in the bottleneck stage i^* is used as a surrogate for minimum makespan value.

$$LBC_{max} = \max_{1 \leq i \leq S}\{\lceil \sum_{j=1}^{N} p_{ij}d_j/M_i \rceil + \sum_{l \neq i}\min_{j}\{p_{lj}\}\} + \sum_{i=1}^{S-1} q_i =$$

$$= \lceil w^* \rceil + \sum_{i \neq i^*}\min_{j}\{p_{ij}\} + \sum_{i=1}^{S-1} q_i \quad (6.14)$$

where $\lceil a \rceil$ is the smallest integer not less than a.

6.1.2 Numerical examples

In this section an illustrative example for a three-stage assembly line is presented. The line (see Fig. 6.1) is made up of $M_1 = 2$ machines in stage $i = 1$, $M_2 = 3$ machines in stage $i = 2$, and $M_3 = 2$ machines in stage $i = S = 3$. The number of buffers ahead of stage 2 and stage 3 are $B_2 = 3$ and $B_3 = 3$, respectively. Transportation times between stages are $q_1 = q_2 = 1$.

The production program consists of $N = 4$ part types and their corresponding production requirements (in number of parts) are $d_1 = 8$, $d_2 = 4$, $d_3 = 2$, $d_4 = 3$. Therefore, the total number of parts to be scheduled is $P = 17$. Assembly times p_{ij} ($i = 1, 2, 3$; $j = 1, 2, 3, 4$) for each stage i and each part type j are given below:

$$p_{11} = 5, \; p_{21} = 3, \; p_{31} = 7,$$
$$p_{12} = 2, \; p_{22} = 4, \; p_{32} = 6,$$
$$p_{13} = 3, \; p_{23} = 6, \; p_{33} = 1,$$
$$p_{14} = 1, \; p_{24} = 4, \; p_{34} = 2.$$

6.1 Flexible assembly line with limited intermediate buffers

Table 6.2. Assembly schedule for the example

Periods	Assignment of parts to machines and buffers												
	Stage 1		Stage 2						Stage 3				
	Machines		Buffers			Machines			Buffers			Machines	
from - to	1	2	1	2	3	1	2	3	1	2	3	1	2
1 - 3	2	2											
4	2	2				2	2						
5	4	3				2	2						
6	1	3	2			2	2	2					
7	1	3	2	4		2	2	2					
8	1	3				2	4	2					
9	1	3			3	2	4	2				2	2
10	1	3				2	4	3				2	2
11	4	4				2	4	3			2	2	2
12	1	1				1	3	3			2	2	2
13-14	1	1	4	4		1	3	3	2	4	2	2	2
15	1	1	4			4	3	3		4		2	2
16	1	1				4	3	4		4		2	2
17	1	1				4	3	4	3	4	1	2	2
18 - 19	1	1		1	1	4	3	4	3	4	1	2	2
20	1	1				1	1	4	3	4	1	2	2
21	1	1				1	1		3	4	1	4	3
22	1	1				1	1		3	4	4	4	1
23	1	1				1		1		4	4	3	1
24 - 25	1	1				1		1	1	1	4	4	1
26	1	1				1			1	1		4	1
27	1								1	1	1	4	1
28	1						1	1	1	1	1	1	1
29 - 30	1						1	1	1		1	1	1
31	1							1	1		1	1	1
32								1	1	1	1	1	1
33				1				1	1	1	1	1	1
34 - 35				1					1	1	1	1	1
36									1	1	1	1	1
37 - 41									1	1	1	1	1
42									1		1	1	1
43 - 48									1			1	1
49												1	1
50 - 55												1	

Analysis of the above data results in finding the bottleneck stage $i^* = 3$ with the average machine workload $w^* = 44$.

The assembly schedule for the example has been obtained using the RITM heuristic with different values of parameters a_1 and a_2, where $a_1 \in \{0.00, 0.25, 0.50, 0.75, 1.00\}$ and $a_2 \in \{1.00, 1.01, \ldots, 1.99, 2.00\}$. The best solution which has been obtained for all values of a_1 and for $1.00 \leq a_2 \leq 1.10$ or $1.46 \leq a_2 \leq 1.99$ yields a schedule of length $C_{max} = 54$ with the loading sequence

$$[4, 4, 4, 2, 2, 2, 1, 2, 3, 3, 1, 1, 1, 1, 1, 1, 1].$$

126   6. Production Scheduling in Flexible Assembly Lines

For all remaining values of parameter a_2 the algorithm yields a schedule with the makespan $C_{max} = 55$ and the loading sequence

$$[2,2,2,2,4,3,1,3,4,4,1,1,1,1,1,1,1].$$

A complete assembly schedule for the latter case is shown in Table 6.2, which has been obtained using the RITM heuristic with parameters $a_1 = 0.5$, $a_2 = 1.2$.

In Table 6.2, the assignment of parts to machines and buffers in each stage is indicated for every period of unit time duration. Numbers of the first and the last period of each assignment are given in the first column of the table. The minimum values of total weighted idle time $\bar{t}_{j_k}$ for every iteration $k = 1, \ldots, 17$ are following:

$$7.6, 7.6, 3.5, 0.0, 0.0, 0.0, 0.0, 1.7, 1.0, 1.0, 0.0, 0.0, 1.2, 0.58, 2.9, 2.9, 3.5$$

In order to better illustrate the solution procedure some of the computations performed in iteration number 8 of the RITM algorithm are shown below as an example.

The partial loading sequence obtained during the first 7 iterations is $[j_1, j_2, \ldots, j_7] = [2, 2, 2, 2, 4, 3, 1]$. The partial schedule for $k = 7$ parts selected so far leads to the following values of times X_{bik} ($b = 1, \ldots, B_i$; $i = 2, 3$) and Y_{mik} ($m = 1, \ldots, M_i$; $i = 1, 2, 3$) at which, respectively buffers and machines are available for assignment,

in stage 1: $Y_{117} = 10$, $Y_{217} = 7$;
in stage 2: $X_{127} = 11$, $X_{227} = 7$, $X_{327} = 9$, $Y_{127} = 14$, $Y_{227} = 11$, $Y_{327} = 15$;
in stage 3: $X_{137} = 20$, $X_{237} = 20$, $X_{337} = 21$, $Y_{137} = 22$, $Y_{237} = 28$.

The resulting earliest availability times: x_{ik} for buffer $b(i, 8)$ ahead of stage i ($i = 2, 3$), and y_{ik} for machine $m(i, 8)$ in stage i ($i = 1, 2, 3$) are shown below $x_{27} = 7$, $x_{37} = 20$, $y_{17} = 7$, $y_{27} = 11$, $y_{37} = 22$.

Therefore, the best route along the assembly line is the following sequence of machines and buffers available at the earliest times:

$$[m(1,8), b(2,8), m(2,8), b(3,8), m(3,8)] = [2, 2, 2, 1, 1].$$

In iteration 8, part j_8 for loading at position $k = 8$ has to be selected from among 10 remaining parts of 3 types $j = 1, 3, 4$. A complete assembly schedule obtained along the best route for each of those 3 part types is shown in Fig. 6.3, respectively for part type 1 (Fig. 6.3(a)), part type 3 (Fig. 6.3(b)), and part type 4 (Fig. 6.3(c)). The corresponding total weighted idle times are $\bar{t}_1 = 2.9$, $\bar{t}_3 = 1.7$, $\bar{t}_4 = 4$, and hence the minimum is $\bar{t}_3 = 1.7$. Therefore, part type $j_8 = 3$ is selected for loading at position $k = 8$, and in iteration 8 its complete assembly schedule shown in Fig. 6.3(b) is added to the cumulative partial schedule of the first 7 parts.

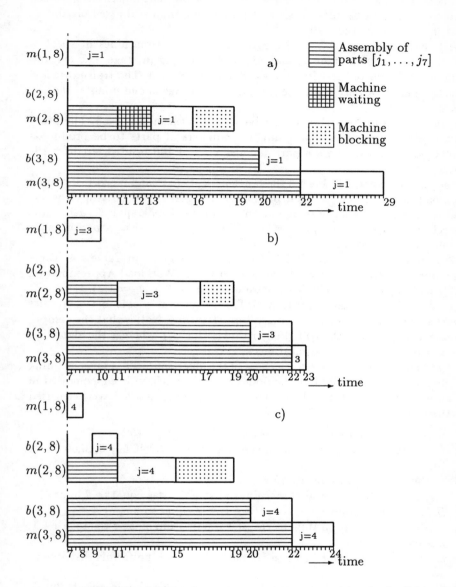

Fig. 6.3. Assembly schedules for part types $j = 1, 3, 4$ waiting for loading

128   6. Production Scheduling in Flexible Assembly Lines

Computational experiments. Results of some computational experiments with the algorithm RITM are reported below. The algorithm has been used to compute schedules for six test problems from a real assembly line at IBM, presented in [128, 129].

The line is made up of $S = 3$ stages with $M_1 = 2$ machines in stage 1, $M_2 = 3$ machines in stage 2, and $M_3 = 3$ machines in stage 3. Each machine has a buffer of capacity one, so that $B_2 = B_3 = 3$. The transportation time to transfer a part from one machine to another is one minute, that is, $q_i = 1$, $i = 1, 2$.

The input data for the IBM test problems are given in Table 6.3. The problems represent production programs (in numbers of parts to be produced) for six typical days, where a part denotes a set of 100 printed circuit cards of one type.

In the last two rows of Table 6.3 the bottleneck stage i^* and its corresponding workload w^* are given for each production program. The bottleneck workload is next used as a lower bound on the optimal makespan to determine the relative error $\epsilon = (C_{max} - w^*)/w^*$. The error is used as a performance measure of the algorithm.

Solution results for the IBM test problems are given in Table 6.4. Table 6.4 reports the comparison of the algorithm WLA (Workload Approximation [129]) and algorithm RITM (with parameters $a_1 = 0.5$, $a_2 = 1.2$).

One may observe that algorithm RITM in general yields better results (shorter schedules) and requires much less CPU time. Notice that the computational complexity of the RITM algorithm is $O(P^2 S)$, whereas the algorithm WLA requires $O(P^3 S)$ computations.

In order to evaluate the effectiveness of the algorithm RITM for scheduling the flexible assembly line, 1000 random test problems have been generated in 10 groups, each of 100 problems. The following parameters have been used for the test problems:

(1) N, number of part types, was equal to 5,8,10,12, or 15;
(2) P, the total number of parts, was equal to 10,20,30,40,50,60,70,80,90, or 100;
(3) S, the number of assembly stages, was equal to 3,4, or 5;
(4) M_i, the number of parallel machines in stage i, was equal to 2,3, or 4;
(5) B_i, number of buffers ahead of stage i, was equal to 2,3, or 4;
(6) q_i, the transportation time to transfer a part from stage i to stage $i+1$, was equal to 1;
(7) p_{ij}, the assembly time in stage i of part type j, was uniformly distributed over [0,200];
(8) d_j, the demand for part type j, was uniformly distributed over [1,30].

Basic input data for each group of the test problems are shown in Table 6.5.

For each random test problem the relative error $\epsilon = (C_{max} - LBC_{max})/LBC_{max}$ (average deviation from the lower bound) was computed as a performance measure of the RITM heuristic (with parameters

6.1 Flexible assembly line with limited intermediate buffers 129

Table 6.3. Input Data for the IBM test problems

Part type j	Assembly Times p_{ij} (in minutes) Stage number:			Production Requirements d_j (in parts) Problem:					
	$i=1$	$i=2$	$i=3$	1	2	3	4	5	6
1	39	11	14	12	-	-	-	-	-
2	13	28	54	1	-	-	-	-	-
3	22	56	60	26	-	-	14	23	20
4	234	39	0	-	-	-	2	-	-
5	39	25	80	-	6	7	4	-	1
6	13	70	54	-	14	20	16	-	-
7	143	66	0	1	4	-	-	3	1
8	0	28	14	7	-	-	-	-	-
9	26	39	74	-	6	-	-	-	5
10	18	59	34	-	4	-	-	-	-
11	22	70	40	4	-	-	-	-	-
12	13	70	54	-	4	-	-	-	-
13	61	46	34	-	-	11	-	14	3
Bottleneck stage i^*				2	2	2	2	1	3
Bottleneck workload w^*				720	715	694	694	895	584

Table 6.4. Results for the IBM test problems

Problem	Bound w^*	Algorithm WLA[129]			Algorithm RITM		
		C_{max}	ϵ	$CPU^{(a)}$	C_{max}	ϵ	$CPU^{(b)}$
1	720	907	25.97%	3.4	837	16.25%	0.16
2	715	972	35.94%	1.7	870	21.68%	0.11
3	694	885	27.52%	1.8	838	20.75%	0.05
4	694	940	35.45%	1.6	922	32.85%	0.05
5	895	964	7.71%	1.8	999	11.62%	0.05
6	584	686	17.46%	1.2	674	15.41%	0.06

(a) CPU seconds on an IBM 3081
(b) CPU seconds on a PC/AT for RITM with $a_1 = 0.5$, $a_2 = 1.2$

Table 6.5. Input data and results for random test problems

Group of problems	N	P	S	Number of machines and number of buffers					$\bar{\epsilon}$ [%]	σ
				M_1	B_2/M_2	B_3/M_3	B_4/M_4	B_5/M_5		
1	5	10	3	3	3/3	3/3	-	-	56	19
2	5	20	3	3	3/2	3/3	-	-	20	10
3	8	30	3	2	3/3	2/3	-	-	17	8
4	8	40	3	2	2/2	2/2	-	-	12	7
5	10	50	4	3	3/3	3/3	3/3	-	21	8
6	10	60	4	3	3/3	3/2	2/3	-	18	10
7	12	70	4	2	3/3	3/2	3/3	-	16	9
8	12	80	5	2	2/2	2/2	2/2	2/2	17	8
9	15	90	5	3	3/3	3/3	3/3	3/3	21	8
10	15	100	5	4	4/4	4/4	4/4	4/4	23	9

N – number of part types, P – total number of parts, S – number of stages, $\bar{\epsilon}$ – average value of the relative error $\epsilon = (C_{max} - LBC_{max})/LBC_{max}$ for a group of 100 test problems, σ – standard deviation of the relative error.

$a_1 = 0.5$, $a_2 = 1.2$). In the last two columns of Table 6.5, mean value $\bar{\epsilon}$ of the relative errors and its standard deviation σ are reported for each group of 100 test problems.

The average CPU time (in seconds) on a PC/AT for one problem in each group of 100 test problems has been 0.07, 0.09, 0.19, 0.20, 0.38, 0.43, 0.62, 0.75, 1.04, 1.16, respectively for problems in group 1,2,3,4,5,6,7,8,9,10.

The computational experiments have also been performed to evaluate how sensitive are the results obtained to the values of parameters a_1 and a_2. A systematic "what–if" analysis has been conducted to solve the IBM test problems using the RITM heuristic with $a_1 \in \{0.0, 0.1, 0.2, \ldots, 0.9, 1.0\}$ and $a_2 \in \{1.00, 1.01, \ldots, 1.99, 2.00\}$. The results of the simulation study have indicated that the best solution is almost insensitive to a_1 and much more dependent on a_2. However, for a given problem, the best solution can be obtained for many different values of a_2. Although the best choice of a_1 and a_2 is dependent on the problem instance, on an average parameters $a_1 = 0.5$ and $a_2 = 1.2$ lead to the best results. Therefore, these values have been selected to solve by a single-pass of RITM both the IBM test problems and the random problems presented in this section.

Notice that the RITM algorithm constructs complete assembly schedule for each new individual part selected for loading, based on the cumulative partial schedule obtained so far. Therefore, the approach presented can also be applied in a dynamic scheduling environment for on-line use.

6.2 Flexible assembly line with no in-process buffers

This section presents a single-pass heuristic algorithm for the scheduling of parts through a flexible assembly line with no in-process buffers [104]. Since there are no buffers between the stages, intermediate queues of parts waiting in the system for their next operations are not allowed.

A flexible assembly line with no intermediate buffers represents a special type of traditional flowshop with no store constraints [19] in which there is only one machine in each stage, and every part visits every machine. A part completed on a machine blocks this machine until the next machine is available for assignment.

The other extreme case is the no-wait flowshop [19], where a part once started on the first stage must be continuously processed through completion at the last stage without interruption. Then obviously no buffer storage is required in this case. However, in addition machine blocking is not allowed. Therefore, the no-wait constraint appears to be more severe than the no-store constraint, e.g., [38].

The literature on scheduling flowshops with no intermediate buffers is mostly restricted to analysis of some special cases, such as two-machine flowshop ([31]), or two-stage, multi-machine flowshop [90] for which various optimizing and approximation algorithms have been proposed. Several heuristics

for scheduling flowshops with limited buffers are presented in [65].
Similarly, the research on the development of scheduling algorithms for the flowshops with parallel machines and unlimited buffers aims at achieving simple heuristics for some special cases, which obtain good solutions in reasonable computing times, for example, see [44, 76, 77, 118]. A branch and bound algorithm for the general flowshop with parallel machines is developed in [25]. However, the computational effort required renders it impractical to solve large-sized problems.

The problem of scheduling a flexible assembly line with no in-process buffers and its mathematical formulation are presented below (notation used to formulate the problem and the algorithm is introduced in Table 6.1).

The flexible assembly line under study consists of $S \geq 2$ assembly stages in series, where each stage i ($i = 1, \ldots, S$) is made up of $M_i \geq 1$ identical parallel machines (see Fig. 6.4). The system produces N different part types. Let $p_{ij} \geq 0$ be the assembly time in stage i of part type j ($j = 1, \ldots, N$), and let q_i be the transportation time required to transfer part from stage i to stage $i + 1$.

A part completed in stage i is transferred to an available machine in the next stage $i+1$ (or another downstream stage depending on the part assembly route). If all M_{i+1} machines in stage $i+1$ are occupied, then machine in stage i is blocked by the completed part until a machine in stage $i + 1$ becomes available.

Given is a production program $(d_1, \ldots, d_N)$, where d_j ($j = 1, \ldots, N$) is the required number of parts of type j. The problem objective is to determine an assignment of all $P = \sum_{j=1}^{N} d_j$ parts to machines in each stage over a scheduling horizon in such a way as to complete the production program in a minimum time, that is, to minimize the makespan $C_{max} = \max_{1 \leq i \leq S}(c_{iP})$, where c_{iP} denotes the completion time in stage i of the last Pth part.

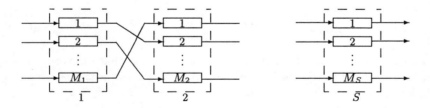

Fig. 6.4. A flexible assembly line with no intermediate buffers

Similarly as for the line with intermediate buffers, the assembly schedule for all P parts is specified by the loading sequence $[j_1, j_2, \ldots, j_P]$ in which the parts enter the line as well as the times required for detailed scheduling of each individual part. In particular, for each part j_k entering the line at position k ($k = 1, \ldots, P$) its start time s_{ik}, completion time c_{ik}, and release time r_{ik} in each stage i should be determined. These variables satisfy the formulae presented below.

The earliest start time of part j_k in stage i is

$$s_{ik} = \max\{r_{i-1,k} + q_{i-1},\ y_{i,k-1}\};\ i = 2,\ldots,S \qquad (6.15)$$
$$s_{1k} = y_{1,k-1}$$

i.e., j_k cannot start assembly in stage i before it has been released from stage $i-1$ and then transferred to stage i, and a machine is available in this stage. Assembly of j_k at the first stage starts when a machine is available after the first $k-1$ parts have been scheduled.

Since no preemption of assembly on a machine is allowed, s_{ik} and c_{ik} for each part j_k and stage i are connected via:

$$c_{ik} = s_{ik} + p_{ij_k};\ i = 1,\ldots,S \qquad (6.16)$$

The release time of part j_k from stage i is

$$r_{ik} = \max\{c_{ik};\ y_{i+1,k-1} - q_i\},\ i = 1,\ldots,S-1 \qquad (6.17)$$
$$r_{Sk} = c_{Sk}$$

i.e., j_k cannot be released from stage i before it has been completed in this stage and a machine is available in stage $i+1$ after the first $k-1$ parts have been scheduled and j_k has been transferred between the two stages.
For the last stage, the release time and completion time are equal.

The earliest available time for stage i after the first k parts have been scheduled is

$$y_{ik} = \min_{1 \leq m \leq M_i}\{Y_{mik}\};\ i = 1,\ldots,S \qquad (6.18)$$

where the earliest available times Y_{mik} for machines in stage i are calculated iteratively as follows

$$Y_{mik} = \begin{cases} Y_{mik-1} & if\ m \neq m(i,k) \\ r_{ik} & if\ m = m(i,k) \end{cases} \qquad (6.19)$$

The formula (6.19) was derived under the assumption that in every stage i part j_k is assigned to the machine $m(i,k)$ with the earliest available time (for additional interpretation of the variables introduced, see Fig. 6.5).

In order to minimize the makespan C_{max}, the RITM-NS heuristic presented in the next section assigns parts to the earliest available machines and in every iteration selects the part which introduces the lowest idle times along its assembly route.

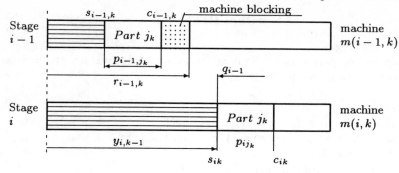

Fig. 6.5. A partial schedule for part j_k

6.2.1 Scheduling algorithm

The algorithm presented in this section, called *Route Idle Time Minimization - No Store* (RITM-NS), is a special variant of the RITM heuristic designed for scheduling flexible assembly lines with limited intermediate buffers (see Sect. 6.1.1.).

All part types are initially ordered according to nonincreasing total assembly times

$$p_j = \sum_{i=1}^{S} p_{ij} \qquad (6.20)$$

During every iteration a part for loading into the system is chosen as well as its complete assembly schedule is determined. The decisions in every iteration are made based on the complete assembly schedule determined for each part type waiting for loading.

Given a cumulative partial schedule for $(k-1)$ parts selected so far, first the best route along the line is found as a sequence

$$[m(1,k), m(2,k), \ldots, m(S,k)] \qquad (6.21)$$

of S machines (one in every stage) with the earliest available times. For each part type waiting for entering the line the complete assembly schedule is determined along the best route. To evaluate the assembly schedule for each part type considered for loading, the total duration of machine idle time along the route is determined

$$t_j = \sum_{i=1}^{S} t_{ij} \qquad (6.22)$$

Finally, the part with the smallest total machine idle time t_j is selected for loading and its complete assembly schedule is added to the cumulative partial schedule obtained so far.

Description of the algorithm RITM-NS is given below (J denotes the set of part types loaded in the required number of parts).

Algorithm RITM-NS

STEP 0. Starting
1. Order all part types according to nonincreasing values of total assembly times p_j, (6.20), that is
$p_1 \geq p_2 \geq \cdots \geq p_N$.
2. Set:
$Y_{mi0} = 0; \; m = 1\ldots,M_i, \; i = 1,\ldots,S;$
$y_{i0} = 0; \; i = 1,\ldots,S; \; m(i,1) = 1; \; i = 1,\ldots,S;$
$J = \emptyset, \; k = 1.$

STEP 1. Selection of a part type for loading
1. For each part type $j \notin J$ waiting for loading determine start time s_{ij}, (6.15), completion time c_{ij}, (6.16), release time r_{ij}, (6.17), and duration t_{ij}, (6.8) of machine idle time, in every stage i on machine $m(i,k)$ with the earliest available time y_{ik-1}. Next, determine total machine idle time t_j (6.22).
2. Select for loading such a part type j_k for which the total machine idle time is minimal
$$j_k = \arg\min_{j \notin J}\{t_j\}$$

To break ties, select the part type with the lowest number (i.e., with the largest total assembly time).

STEP 2. Determining complete assembly schedule for the selected part
For the selected part j_k determine complete assembly schedule (times s_{ik}, (6.15) c_{ik}, (6.16) r_{ik}, (6.17)) by assigning it in every stage i to the machine $m(i,k)$ with the earliest available time, y_{ik-1},(6.18), after the first $k-1$ parts have been scheduled.
The assembly schedule for part j_k add to the cumulative partial schedule for the first $(k-1)$ parts.

STEP 3. Checking the state of completion the production program
For $j = j_k$ set $d_j = d_j - 1$. If $d_j = 0$, then set $J = J \bigcup \{j\}$.
If $J = \{1,\ldots,N\}$, then terminate.
Otherwise determine for each machine m $(m = 1,\ldots,M_i, \; i = 1,\ldots,S)$ the earliest available time Y_{mik}, (6.19) after the first k parts have been scheduled.
For each stage i find machine $m(i,k+1) = \arg\min_{1 \leq m \leq M_i}\{Y_{mik}\}$ with the earliest available time y_{ik}, (6.18).
Set $k = k + 1$ and go to *STEP 1*.

In order to determine the computational complexity of the algorithm RITM-NS notice that the algorithm requires P iterations and in every iteration a complete assembly schedule is computed for one part according to the following procedure.
First, the best route is found as a sequence of at most S machines in the successive stages with the earliest available times. This step requires $O(M)$

computations since at most all M machines are considered ($M = \sum_{i=1}^{S} M_i$ is the total number of machines).

Then, for each of the remaining at most N part types, the complete assembly schedule is computed based on the best route found. This step requires $O(NS)$ computations to determine the times for the N flowshop schedules on S machines. The best assembly schedule (and by this a part for loading) with the smallest total idle time is next chosen and added to the cumulative partial schedule for all parts loaded so far.

Since the above procedure which requires $O(M + NS)$ computations is repeated in each of the P iterations, and the number of part types $N \leq P$, the computational complexity of the algorithm RITM-NS is $O(P^2 S + PM)$.

6.2.2 Numerical examples

In this section an illustrative example for a three-stage assembly line is presented. The line is made up of $M_1 = 2$ machines in stage $i = 1$, $M_2 = 3$ machines in stage $i = 2$, and $M_3 = 2$ machines in stage $i = S = 3$. Transportation times between stages are $q_1 = q_2 = 1$.

The production program consists of $N = 4$ part types and their corresponding production requirements (in number of parts) are $d_1 = 8$, $d_2 = 4$, $d_3 = 2$, $d_4 = 3$. Therefore, the total number of parts to be scheduled is $P = 17$. Assembly times p_{ij} ($i = 1, 2, 3$; $j = 1, 2, 3, 4$) for each stage i and each part type j are given below:

$$p_{11} = 5,\ p_{21} = 3,\ p_{31} = 7,$$
$$p_{12} = 2,\ p_{22} = 4,\ p_{32} = 6,$$
$$p_{13} = 3,\ p_{23} = 6,\ p_{33} = 1,$$
$$p_{14} = 1,\ p_{24} = 4,\ p_{34} = 2.$$

Analysis of the above data results in finding the bottleneck stage $i^* = 3$ with the average machine workload $w^* = 44$. The lower bound (6.14) on makespan is $LBC_{max} = 44 + 1 + 3 + 1 + 1 = 50$.

The assembly schedule for the example has been obtained using the RITM-NS heuristic. The algorithm yields a schedule of length $C_{max} = 52$ with the loading sequence

$$[4, 4, 2, 1, 3, 2, 3, 1, 1, 1, 1, 2, 1, 1, 1, 2, 4]$$

The corresponding values of minimum total idle time t_{j_k}, (6.8), for parts j_k ($k = 1, \ldots, 17$) selected for loading in every iteration are the following:

$$9, 9, 4, 3, 2, 3, 1, 1, 3, 5, 7, 7, 7, 9, 9, 11, 12.$$

A complete assembly schedule is shown in Table 6.6, where the assignment of parts to machines in each stage is indicated for every period of unit time duration. Numbers of the first and the last period of each assignment are given in the first column of Table 6.6.

Table 6.6. Assembly schedule

Periods from-to	Assignment of parts to machines						
	Stage 1 Machine		Stage 2 Machine			Stage 3 Machine	
	1	2	1	2	3	1	2
1	4	4					
2	2	1					
3	2	1	4	4			
4	3	1	4	4			
5-6	3	1	4	4	2		
7	2	3			2		
8	2	3	1	3	2	4	4
9	1	3	1	3		4	4
10	1	1	1	3	2	2	
11	1	1	3	3	2	2	
12-14	1	1	3	3	2	2	1
15	1	1	3	1	2	2	1
16	1	1	3	1	1	3	1
17	1	1	3	1	1	2	1
18	1	1		1	1	2	1
19	2	1			1	2	3
20	2	1	1		1	2	1
21	1	1	1	1	1	2	1
22	1	1	1	1	2	2	1
23-27	1	1	1	1	2	2	1
28-31	2	1	1	1	2	1	1
32	2	4	1	1	2	1	1
33	2	4	1	1	1	1	1
34	2	4	1	1	1	1	2
35-37		4	2	1	1	1	2
38-39			2	1	1	1	2
40-42			2	4	1	1	1
43-45			2	4		1	1
46				4		1	1
47-49				4		1	2
50						1	2
51-52						4	2

In order to better illustrate the solution procedure some of the computations performed in iteration number 10 of the RITM-NS algorithm are shown below as an example.

The partial loading sequence obtained during the first 9 iterations is $[j_1, \ldots, j_9] = [4, 4, 2, 1, 3, 2, 3, 1, 1]$. The partial schedule for $k = 9$ parts selected so far leads to the following values of the earliest available times Y_{mik} ($m = 1, \ldots, M_i$; $i = 1, 2, 3$) for machines,
in stage 1: $Y_{1,1,9} = 13$, $Y_{2,1,9} = 14$;
in stage 2: $Y_{1,2,9} = 17$, $Y_{2,2,9} = 18$, $Y_{3,2,9} = 21$;
in stage 3: $Y_{1,3,9} = 29$, $Y_{2,3,9} = 26$.

The resulting earliest available times y_{ik} for machine $m(i, 10)$ in stage i ($i = 1, 2, 3$) are $y_{1,10} = 13$, $y_{2,10} = 17$, $y_{3,10} = 26$.

Therefore, the best route along the assembly line is the following sequence of machines available at the earliest times:

$$[m(1,10), m(2,10), m(3,10)] = [1,1,2].$$

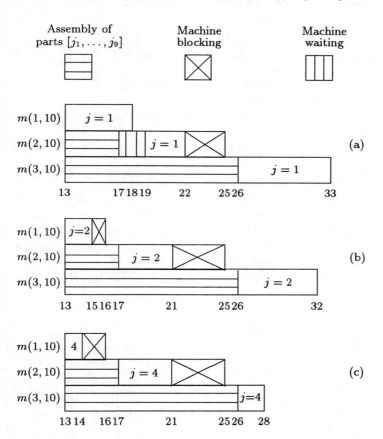

Fig. 6.6. Assembly schedules for part types $j = 1, 2, 4$ waiting for loading

In iteration 10, part j_{10} for loading at position $k = 10$ has to be selected from among 8 remaining parts of 3 types $j = 1, 2, 4$. A complete assembly schedule obtained along the best route for each of those 3 part types is shown in Fig. 6.6, respectively for part type 1 (Fig. 6.6(a)), part type 2 (Fig. 6.6(b)), and part type 4 (Fig. 6.6(c)). The corresponding total idle times are $t_1 = 5$, $t_2 = 5$, $t_4 = 6$, and the minimum is $t_1 = t_2 = 5$. Therefore, part type $j_{10} = 1$ with the least total idle time introduced and the largest total assembly time is selected for loading at position $k = 10$, and in iteration 10 its complete assembly schedule shown in Fig. 6.6(a) is added to the cumulative partial schedule of the first 9 parts.

Computational experiments. In order to evaluate the effectiveness of the algorithm RITM-NS for scheduling the flexible assembly line with no in-process buffers, 1000 random test problems have been generated in 10 groups, each of 100 problems. The following parameters have been used for the test problems:

(1) N, number of part types, was equal to 5,8,10,12, or 15;
(2) P, the total number of parts, was equal to 10,20,30,40,50,60,70,80,90, or 100;
(3) S, the number of assembly stages, was equal to 3,4, or 5;
(4) M_i, the number of parallel machines in stage i, was equal to 2,3, or 4;
(6) q_i, the transportation time to transfer a part from stage i to stage $i+1$, was equal to 1;
(7) p_{ij}, the assembly time in stage i of part type j, was uniformly distributed over [0,200];
(8) d_j, the demand for part type j, was uniformly distributed over [1,30].

Basic input data for each group of the test problems are shown in Table 6.7.

For each random test problem the relative error $\epsilon = (C_{max} - LBC_{max})/LBC_{max}$ (average deviation from the lower bound) was computed as a performance measure of the RITM-NS heuristic. In the last three columns of Table 6.8, mean value $\bar{\epsilon}$ of the relative errors, its standard deviation σ and average CPU time, $\overline{CPU}$ are reported for each group of 100 test problems.

Table 6.7. Input Data and Results for Random Test Problems

Group of problems	N	P	S	Number of machines M_i					$\bar{\epsilon}$ [%]	σ	$\overline{CPU}$ [s]
				M_1	M_2	M_3	M_4	M_5			
1	5	10	3	3	3	3	-	-	22.36	13.58	0.04
2	5	20	3	3	2	3	-	-	10.39	8.32	0.06
3	8	30	3	2	3	3	-	-	7.48	9.66	0.12
4	8	40	3	2	2	2	-	-	7.03	6.00	0.13
5	10	50	4	3	3	3	3	-	12.08	7.85	0.26
6	10	60	4	3	3	2	3	-	6.51	6.01	0.32
7	12	70	4	2	3	2	3	-	7.04	6.96	0.44
8	12	80	5	2	2	2	2	2	10.04	6.72	0.54
9	15	90	5	3	3	3	3	3	14.21	8.27	0.79
10	15	100	5	4	4	4	4	4	19.16	7.53	0.88

N – number of part types, P – total number of parts, S – number of stages,
$\bar{\epsilon}$ – average value of the relative error $\epsilon = (C_{max} - w^*)/w^*$ for a group of 100 test problems,
σ – standard deviation of the relative error, $\overline{CPU}$ – average CPU time

The results of the computational experiments have indicated that the RITM-NS heuristic achieves good solutions in very short CPU run time. The maximum error for 1000 random test problems never exceeded 28% and the average error was not greater than 20%. The computation time required on a PC/AT was not greater than one second for the medium-sized problems that can be encountered in an industrial practice.

The algorithms presented in this chapter for production scheduling in flexible assembly lines achieves good solutions in very short CPU run time.

The algorithms RITM and RITM-NS are single-pass heuristics, in which in every iteration a complete assembly schedule is constructed for each new individual part selected for loading, based on the cumulative partial schedule obtained for the parts that had already been loaded into the line. The above features indicate that the approach presented can also be applied for on-line use in a dynamic scheduling environment. Performance comparison of the RITM and RITM-NS heuristics indicate that better results (in terms of shorter makespans obtained and CPU times required) are achieved for scheduling flexible assembly lines with no in-process buffers, in which the machine blocking phenomenon may occur more frequently. As a result the total idle time along an assembly route may increase which further restricts the selection of a part type for loading and makes the solution space smaller [98, 101, 104].

6.3 Global lower bounds for flexible flow lines with unlimited buffers

Determination of the optimal makespan schedule for flexible assembly lines may be practically impossible even for small sized problems. Therefore, a strong lower bound should be determined to estimate the optimal makespan and evaluate the quality of heuristic procedures which can generate good makespan solutions. In this section a global makespan lower bound is presented for the flexible flow line with unlimited intermediate buffers. In such a line unlimited queueing is allowed. However, the lower bound procedure presented for that type of line can also be used to evaluate the scheduling heuristics for the flexible assembly lines with intermediate constraints.

The procedure for developing a global makespan lower bound for the flexible flow line with unlimited buffers involves the determination of a stage-based lower bound, see [89]. Let LB_i be the stage-based makespan lower bound with respect to stage i ($i = 1, \ldots, S$), and LB_0 – the total processing time of the longest duration product

$$LB_0 = \max_j \{p_j\} = \max_j \left\{ \sum_{i=1}^{S} p_{ij} \right\} \qquad (6.23)$$

The global lower bound can be denoted by LBC_{max} where

$$LBC_{max} = \max \left\{ LB_0, \max_{1 \leq i \leq S} LB_i \right\} \qquad (6.24)$$

A brief description of the mechanism of the bounds proposed is as follows. In an optimal schedule each machine m ($m = 1, \ldots, M_i$) in stage i processes a subset of products J_{im}. The maximum of the completion times of J_{im} can be no less than the completion time of the last product to be processed by machine m. In other words, the completion time of the last product to be processed by machine m establishes a machine-based bound on the products in J_{im}. The maximum of the machine-based bounds over all M_i machines at a stage i establishes the stage-based makespan lower bound on all N products.

Before the last product of machine m in stage i can exit the line, the first product in J_{im} must arrive at that machine, all the products in J_{im} must be processed by machine m, and the last product to be processed by machine m must be processed through each of the remaining stages. Since it is not known exactly which products are assigned to each J_{im}, one cannot determine the maximum of the machine-based bounds in this set. However, it is known that this maximum must be greater than or equal to the average of the M_i machine-based bounds at this stage.

The new stage-based lower bound LB_i with respect to each stage i is based on this concept of finding the average of the M_i machine-based bounds ([89])

$$LB_i = \frac{1}{M_i}\left(\sum_{m=1}^{M_i} LSA(i,m) + \sum_{j=1}^{N} p_{ij} + \sum_{m=1}^{M_i} RSA(i,m)\right) \qquad (6.25)$$

In the case of integer processing times, LB_i should be rounded up to the nearest integer to obtain a lower bound value.

The LS-values refer to the left-side total processing times before stage i and the RS-values are the right-side total processing times after stage i. Similarly, the LSA i RSA lists are the left-side total processing times and the right-side total processing times arranged in ascending order.

For any stage i, the M_i smallest left-side processing times are simply $LSA(i,1), \ldots, LSA(i,M_i)$, where $LSA(i,1) \leq LSA(i,2) \leq \cdots \leq LSA(i,M_i)$. The M_i smallest right-side processing times are similarly, $RSA(i,1), \ldots, RSA(i,M_i)$, where $RSA(i,1) \leq RSA(i,2) \leq \cdots \leq RSA(i,M_i)$.

The left-side and right-side sums of the processing times of product j, respectively before stage i and after stage i are calculated as follows

$$LS(i,j) = \begin{cases} \sum_{i'=1}^{i-1} p_{i'j} & i > 1 \\ 0 & i = 1 \end{cases} \qquad (6.26)$$

$$RS(i,j) = \begin{cases} \sum_{i'=i+1}^{S} p_{i'j} & i < S \\ 0 & i = S \end{cases} \qquad (6.27)$$

6.3.1 Numerical example

Consider an example problem for a $S = 3$ stage flexible flow line made up of 2 machines in each stage, i.e., $M_1 = M_2 = M_3 = 2$, in which $N = 5$ products

6.3 Global lower bounds for flexible flow lines with unlimited buffers

Table 6.8. Processing times for the example problem

Product	Processing times p_{ij}		
j	Stage $i=1$	Stage $i=2$	Stage $i=3$
1	3	5	9
2	7	1	4
3	2	7	4
4	8	2	2
5	6	3	7

are to be processed. Table 6.8 represents the processing times of the various products at the different stages.

Before the bounds are calculated, examine the schedule which loads the machines in product order $[1, 2, 3, 4, 5]$. If the products are loaded in a non-delay manner using the finish order on the preceding stage to break ties, the makespan obtained is 24 time units [89].

In order to evaluate how good this schedule is, the global lower bound will be used. For the example problem, $LB_0 = 17$ which is the total processing time for product $j = 1$. The sum of the product times in stage 1 is 26. For stage 2 the sum is 18 and for stage 3 the sum is 26. The stage-based lower bounds can now be calculated.

For LB_1 there are no left-side processing times. Therefore, only M_1 smallest right-side processing times need be computed. The two smallest right-side processing times are 4 and 5 corresponding to products 4 and 2, respectively. Thus,

$$LB_1 = \frac{1}{2}(26 + (4+5)) \geq 18$$

Note that LB_1 is rounded since the example contains integer processing times p_{ij}.

Concerning stage 2, the M_2 smallest left-side processing times are 2 and 3 corresponding to products 3 and 1, respectively. The two smallest right-side processing times are 2 and 4 corresponding to products 4 and 2, respectively (or 4 and 3). Thus

$$LB_2 = \frac{1}{2}(18 + (2+3) + (2+4)) \geq 15$$

For stage 3, there are no right-side processing times. The M_3 smallest left-side processing times are 8 and 8 corresponding to products 1 and 2, respectively. Thus

$$LB_3 = \frac{1}{2}(26 + (8+8)) = 21.$$

There are now four estimates of the lower bounds and the global bound is

$$LBC_{max} = \max\{17, 18, 15, 21\} = 21$$

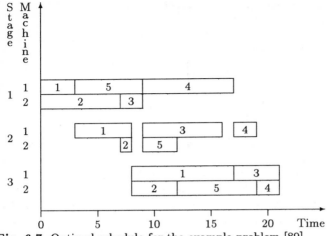

Fig. 6.7. Optimal schedule for the example problem [89]

For this example problem, the lower bound computations accurately predict the optimal makespan of 21, which can be obtained for the input sequence [1, 2, 5, 3, 4], as depicted in Fig. 6.7 [89].

The global lower bound presented in this section for the flexible flow line with unlimited buffers can also be used to evaluate the quality of the heuristics RITM and RITM-NS proposed for scheduling flexible assembly lines with intermediate constraints.

6.4 Just-in-time scheduling of flexible assembly lines

Flexible assembly lines are used to manufacture many final products without holding large inventories. Each product assembled on a flexible assembly line requires a variety of parts. Often these parts vary from product to product. There should be little variability in the usage of each part from one time period to the next. In Hall [47] this objective is referred to as levelling or balancing the schedule. A balanced schedule minimizes the variability in the production rate of all parts and final products throughout a flexible assembly line. Negligible change-over costs between the products in the flexible assembly lines make it possible to implement *Just-In-Time* (JIT) production method, which requires producing only the necessary products in the necessary quantities at the necessary times. Using JIT methods, it is possible to satisfy customer demands for a variety of products without holding large inventories or incurring large shortages, provided there is sufficient production capacity available.

In conventional production planning and control systems finishing products early is regarded as being at least as desirable as finishing on time. In JIT systems, however, products are being penalized both for being early and for being tardy. Therefore, JIT scheduling problems deal with penalties for both earliness and tardiness, see [56].

JIT assembly line is a pull production system in which any supplying process is initiated only if there is another process that requires the supplying process output (subassembly, part, etc.). Once the production schedule is fixed for the final assembly, the schedules for all other production levels are also inherently fixed. Therefore, it is the final assembly level which is the focus for scheduling. This schedule must be such that the quantity of each part used by the assembly system per unit time is kept as close to constant as possible. In other words, there should be very little variability in the usage of each output by the line from one period to the next. For example, if a part is needed for certain final products, its usage is high when those products are being assembled and low otherwise. Also, products do not have the same assembly time at each station on the line. Some may even have assembly times at certain stations that exceed the predetermined cycle time. If such products are scheduled successively, delays and stoppages may occur.

Minimizing variability in the usage of each part from one period to the next was stated ([74]) as the most important goal of a JIT production system by the management of Toyota, the first company to use JIT methods. This eliminates the possibility of "shock waves" created by sudden increases in part requirements, which may result in shortages, or by sudden decreases, which may create excessive inventories. In [74] the following example of balancing the production of Coronas at Toyota is provided. An eight-hour production shift must produce 250 sedans, 125 hardtops and 125 wagons (i.e., essentially 1 car/minute). The balanced production sequence (at the final level) would be: sedan, wagon, sedan, hardtop, sedan, wagon, sedan, hardtop, etc. In order to construct this type of schedule for problems with more product types, Toyota has developed a local search heuristic called *Goal Chasing Method* [74].

A balanced schedule does not account for the work contents of the various products which might be different and can affect station workloads. If different options (i.e., parts such as engines, transmissions, accelerators, number of doors) required by the different variants do affect station workloads, then the schedule for the final assembly should be derived by solving the so-called *car sequencing* problem. For a combined balanced scheduling/car sequencing problem, see [30].

6.4.1 Final assembly scheduling

Consider a flexible assembly line which produces n final product types. Let d_j be the demand for product type j $(j = 1, \ldots, n)$ to be produced during a specified time horizon. Assume that each product takes a unit of time (cycle time) to be produced so that the entire time horizon consists of $T = \sum_{j=1}^{n} d_j$ unit time periods.

Let $r_j = d_j/T$ be the average demand for product type j per unit time. The scheduling objective is to keep the proportion of the cumulative production of product type j to the total production as close to r_j as possible.

Let x_{jt} be the total cumulative production of product type j in periods 1 through t. The problem objective is to minimize the sum of the absolute values of differences between cumulative production and cumulative average demand of various product types over time. The JIT scheduling problem for final assembly is formulated below ([57, 58, 70]).

Scheduling final assembly

Minimize
$$\sum_{j=1}^{n} \sum_{t=1}^{T} |x_{jt} - r_j t| \qquad (6.28)$$

subject to

$$\sum_{j=1}^{n} x_{jt} = t; \qquad t = 1, \ldots, T \qquad (6.29)$$

$$0 \leq x_{jt+1} - x_{jt} \leq 1; \quad j = 1, \ldots, n,\ t = 1, \ldots, T-1 \qquad (6.30)$$

$$x_{jt} \geq 0,\ \text{integer}; \qquad j = 1, \ldots, n,\ t = 1, \ldots, T \qquad (6.31)$$

The objective function (6.28) can be interpreted as one minimizing total cost of inventories and shortages of various product types over time. This is because $(x_{jt} - r_j t)$, when positive (negative), represents excess (under) cumulative production of product type j when compared to its cumulative average demand through the tth period, and therefore, its inventory (shortage) in period t. Constraint (6.29) ensures that exactly t products are produced during the first t periods. Constraints (6.30) and (6.31) ensure that at most one product can be produced in each period. Note that the constraints $x_{jT} \leq d_j$ ($j = 1, \ldots, n$) are not included as these would be satisfied by any optimal solution.

The sequencing problem as an assignment problem. The final assembly JIT scheduling problem can be reduced to simple assignment problem [57, 58]. Such a reduction enables an optimal schedule for the problem to be computed in time that is polynomial in the total number of products to be assembled.

An intermediate step leading to the reduction to an assignment problem is its reformulation into an equivalent problem which decomposes the original problem into n single product type problems with an appropriate linking constraint.

Let t_{jk} be the period in which the kth unit of product type j is produced. The following problem is equivalent to the original final assembly scheduling problem ([58]).

6.4 Just-in-time scheduling of flexible assembly lines

Selection of assembly periods

Minimize

$$\sum_{j=1}^{n} \left(\sum_{t=0}^{t_{j1}-1} |0 - r_j t| + \sum_{t=t_{j1}}^{t_{j2}-1} |1 - r_j t| + \ldots + \sum_{t=t_{jd_j}}^{T} |d_j - r_j t| \right) \quad (6.32)$$

subject to

$$t_{jk+1} \geq t_{jk} + 1; \quad j = 1, \ldots, n, \ k = 1, \ldots, d_j \quad (6.33)$$

$$1 \leq t_{jk} \leq T; \quad j = 1, \ldots n, \ k = 1, \ldots, d_j \quad (6.34)$$

$$t_{jk} \neq t_{j_1 k_1}; \quad (j, k) \neq (j_1, k_1) \quad (6.35)$$

$$t_{jk} \geq 0, \text{ integer}; \quad j = 1, \ldots, n, \ k = 1, \ldots, d_j \quad (6.36)$$

The only linking constraint is (6.35). Note that (6.35) is not in the standard integer programming format. It specifies that only one product type can be produced in each period.

The coefficients $|k - r_j t|$ appearing in the objective function, (6.32) are plotted in Fig. 6.8 ([58]) for $k = 0, 1, 2, 3$, $d_j = 3$, $T = 17$, and thus $r_j = 3/17$. For example, $|1 - r_j t|$ in this figure begins with a value of 1 at $t = 0$, decreases as t increases, becomes 0 at $t = 1/r_j = 17/3$, and then increases to the value of 2 at $t = T = 17$.

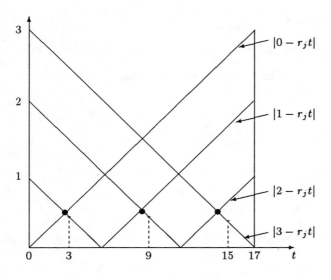

Fig. 6.8. Coefficients $|k - r_j t|$ for $d_j = 3$ and $T = 17$ [58]

6. Production Scheduling in Flexible Assembly Lines

The assembly periods selection problem could be solved on a product-by-product basis, if it were not for constraint (6.35) which is an assignment type constraint.

Before the equivalent assignment problem is presented, define the set $J = \{(j,k) : j = 1, \ldots, n, \ k = 1, \ldots, d_j\}$, so that a pair (j,k) denotes the kth unit of product type j.

Let

$$t_{jk}^* = \lceil (2k-1)/2r_j \rceil; \ k = 1, \ldots, d_j \quad (6.37)$$

be the *ideal period* or position for producing the kth unit of product type j. Period t_{jk}^* is the ideal position for (j,k) to be assigned. The ideal assignment periods are determined as solutions to the n independent problems (6.32) – (6.36) of selection assembly periods independently for each product type in the absence of linking constraints (6.35).

Note that $(2k-1)/2r_j$ represents the point in continuous time where $|k - r_j t|$ and $|k - 1 - r_j t|$ cross one another, i.e., the point satisfying

$$|k - r_j t| = |k - 1 - r_j t| \quad (6.38)$$

The crossing points of the above functions at times $(2k-1)/2r_j$ for $k = 1, 2, 3$ are shown in Fig. 6.8 where dotted lines additionally indicate the corresponding ideal periods t_{jk}^*.

Denote by c_{jkt} the cost of assigning the kth unit of product type j to the period t and introduce the following decision variable

$$z_{jkt} = \begin{cases} 1 & \text{if } (j,k) \text{ is assigned to period } t \\ 0 & \text{otherwise} \end{cases} \quad (6.39)$$

Then, the equivalent assignment problem is formulated below ([57, 58]).

The assignment problem

Minimize

$$\sum_{(j,k) \in J} \sum_{t=1}^{T} c_{jkt} z_{jkt} \quad (6.40)$$

subject to

$$\sum_{(j,k) \in J} z_{jkt} = 1; \quad t = 1, \ldots, T \quad (6.41)$$

$$\sum_{t=1}^{T} z_{jkt} = 1; \quad (j,k) \in J \quad (6.42)$$

$$z_{jkt} \in \{0,1\}; \quad (j,k) \in J, \ t = 1, \ldots, T \quad (6.43)$$

The above assignment problem has an optimal solution, which is both feasible and optimal for the original final assembly scheduling problem if the cost coefficients are defined as follows

6.4 Just-in-time scheduling of flexible assembly lines

$$c_{jkt} = \begin{cases} \sum_{l=t}^{t_{jk}^*-1} e_{jkl} & \text{if } t < t_{jk}^* \\ 0 & \text{if } t = t_{jk}^* \\ \sum_{l=t_{jk}^*}^{t-1} e_{jkl} & \text{if } t > t_{jk}^* \end{cases} \quad (6.44)$$

where

$$e_{jkl} = ||k-r_j l|-|k-1-r_j l|| = \begin{cases} |k-r_j l| - |k-1-r_j l| & \text{if } l < t_{jk}^* \\ |k-1-r_j l| - |k-r_j l| & \text{if } l \geq t_{jk}^* \end{cases} \quad (6.45)$$

Notice that $|k - r_j l|$ represents the inventory or shortage cost in period l if k units of product type j have been produced by period l. The coefficient e_{jkl} represents the excess cost of having k units of product type j, over having $k-1$ units of the same product type, produced by period l.

The coefficient c_{jkt} represents additional cost of assigning the kth unit of product type j to the period t. If $t = t_{jk}^*$, then (j, k) has its ideal position and $c_{jkt} = 0$. If $t < t_{jk}^*$, i.e., the kth unit of product type j is produced too soon, then excess inventory costs e_{jkl} are incurred in periods from $l = t$ to $l = t_{jk}^* - 1$. On the other hand, if $t > t_{jk}^*$, i.e., if the kth unit of product type j is produced too late, then the excess shortage costs e_{jkl} are incurred in periods from $l = t_{jk}^*$ to $l = t - 1$.

It should be noted that the assignment problem is one of the most efficiently solved problems in the operations research literature, e.g., [80].

6.4.2 Numerical example

The assignment problem (6.40) – (6.43) is applied to scheduling three final product types 1, 2 and 3 with demands $d_1 = 2$, $d_2 = 3$ i $d_3 = 5$, respectively. Thus $T = 10$, $r_1 = 0.2$, $r_2 = 0.3$ and $r_3 = 0.5$.

Table 6.9. Excess inventory/shortage costs e_{jkl} [58]

j	$k \backslash l$	1	2	3	4	5	6	7	8	9	10
1	1	0.6	0.2	0.2	0.6	1	1	1	1	1	1
	2	1	1	1	1	1	0.6	0.2	0.2	0.6	1
2	1	0.4	0.2	0.8	1	1	1	1	1	1	1
	2	1	1	1	0.6	0	0.6	1	1	1	1
	3	1	1	1	1	1	1	0.8	0.2	0.4	1
3	1	0	1	1	1	1	1	1	1	1	1
	2	1	1	0	1	1	1	1	1	1	1
	3	1	1	1	1	0	1	1	1	1	1
	4	1	1	1	1	1	1	0	1	1	1
	5	1	1	1	1	1	1	1	1	0	1

The ideal positions t_{jk}^* computed by using the formula (6.37) are as follows ([58]):

148   6. Production Scheduling in Flexible Assembly Lines

Table 6.10. Assignment costs c_{jkt} [58]

j	$k\backslash t$	1	2	3	4	5	6	7	8	9	10
1	1	0.8	0.2	0	0.2	0.8	1.8	2.8	3.8	4.8	5.8
	2	5.8	4.8	3.8	2.8	1.8	0.8	0.2	0	0.2	0.8
2	1	0.4	0	0.2	1	2	3	4	5	6	7
	2	3.6	2.6	1.6	0.6	0	0	0.6	1.6	2.6	3.6
	3	7	6	5	4	3	2	1	0.2	0	0.4
3	1	0	0	1	2	3	4	5	6	7	8
	2	2	1	0	0	1	2	3	4	5	6
	3	4	3	2	1	0	0	1	2	3	4
	4	6	5	4	3	2	1	0	0	1	2
	5	8	7	6	5	4	3	2	1	0	0

$t_{11}^* = 3$ and $t_{12}^* = 8$ for product type 1;
$t_{21}^* = 2$, $t_{22}^* = 5$ and $t_{23}^* = 9$ for product type 2;
$t_{31}^* = 1$, $t_{32}^* = 3$, $t_{33}^* = 5$, $t_{34}^* = 7$ and $t_{35}^* = 9$ for product type 3.

The excess inventory or shortage costs $e_{jkl} = ||k - r_j l| - |k - 1 - r_j l||$ ($j = 1, 2, 3$, $k = 1, \ldots, d_j$, $l = 1, \ldots, T$) are computed by using the formula (6.45) and are shown in Table 6.9. The assignment costs c_{jkt} computed by using the formula (6.44) are shown in Table 6.10.

One may observe (see Table 6.10) that schedules: $3, 2, 1, 3, 3, 2, 3, 1, 2, 3$ and $3, 2, 1, 3, 2, 3, 3, 1, 2, 3$ both have a zero value of the objective function (6.40) and hence are optimal.

6.4.3 Balanced scheduling of a multilevel flexible assembly line

Consider a flexible assembly line made up of S stages (assembly levels), see Fig. 6.9. The highest level (i.e., the final assembly level) is level S, in which various final products are assembled of parts and subassamblies produced at lower levels $1, \ldots, S - 1$.

The number of different part types of level i are denoted by n_i, and d_{ij} represents the demand for part type j of level i ($i = 1, \ldots, S$, $j = 1, \ldots, n_i$). Let g_{ijp} be the amount of part type j of level i required to assemble one unit of final product type p ($p = 1, \ldots n_S$) at level S. For convenience, define $g_{Sjp} = 1$ if $j = p$; otherwise $g_{Sjp} = 0$.

The demand $d_{ij} = \sum_{p=1}^{n_S} g_{ijp} d_{Sp}$ for part type j of level i is directly dependent on the final product demands d_{Sp}, $p = 1, \ldots, n_S$.

Let $D_i = \sum_{j=1}^{n_i} d_{ij}$ be the total part demand of level i. Denote by $r_{ij} = d_{ij}/D_i$ the demand ratio for part type j of level i. For example, $r_{32} = 0.2$ means that 20% of the total output on level 3 is of part type 2 made at this level, and in an ideally balanced schedule exactly every 5th output of level 3 would be part type 2. (Note that $\sum_{j=1}^{n_i} r_{ij} = 1, \forall i = 1, \ldots, S$).

Assume that production of one unit of final product type p ($p = 1, \ldots, n_S$) at level S requires one time period which represents the cycle time. Thus, the entire time horizon T is made up of $D_S = \sum_{j=1}^{n_S} d_{Sj}$ periods, i.e., $T = D_S$.

6.4 Just-in-time scheduling of flexible assembly lines

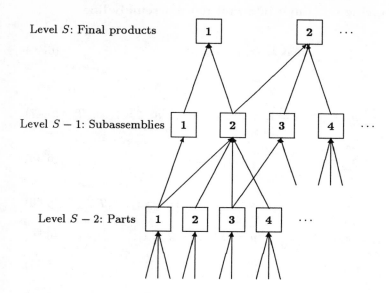

Fig. 6.9. Multilevel assembly line

If exactly t units of various final products have been assembled at level S it means that exactly t periods (cycles) have elapsed.

By the pull nature of JIT systems and from the fact that lower level parts (from levels $i = 1, \ldots, S-1$) are drawn as needed by the final assembly level S, the particular combination of the final products assembled at level S during the first t periods determines the necessary cumulative part production at every other level.

Let x_{ijt} be the necessary cumulative production of part type j of level i during the first t periods, and let $y_{it} = \sum_{j=1}^{n_i} x_{ijt}$ be the total output of level i during the first t periods. (The cumulative production of final level S during the first t periods is $y_{St} = \sum_{p=1}^{n_S} x_{Spt} = t$). The required cumulative production for part type j at level i ($i \leq S-1$) during the first t periods is $x_{ijt} = \sum_{p=1}^{n_S} g_{ijp} x_{Spt}$.

Denote by $\lambda_{ij} \geq 0$ a weighting factor, which reflects the relative importance of balancing the schedule for part type j at level i.

The min-max balanced scheduling problem to select the x_{Spt} ($p = 1, \ldots, n_S$, $t = 1, \ldots, T$) has the following nonlinear integer programming formulation (see [59, 71])

Balancing schedules in a multilevel assembly line

Minimize
$$\max_{i,j,t} \lambda_{ij} |x_{ijt} - r_{ij} y_{it}| \qquad (6.46)$$

subject to

$$x_{ijt} = \sum_{p=1}^{n_S} g_{ijp} x_{Spt}; \quad i=1,\ldots,S,\ j=1,\ldots,n_i, t=1,\ldots,T \qquad (6.47)$$

$$y_{St} = \sum_{p=1}^{n_S} x_{Spt}; \qquad t=1,\ldots,T \qquad (6.48)$$

$$y_{it} = \sum_{j=1}^{n_i} x_{ijt}; \qquad i=1,\ldots,S-1,\ t=1,\ldots,T \qquad (6.49)$$

$$x_{Spt} \geq x_{Spt-1}; \qquad p=1,\ldots,n_S,\ t=1,\ldots,T \qquad (6.50)$$

$$\sum_{p=1}^{n_S} x_{Spt} = t; \qquad t=1,\ldots,T \qquad (6.51)$$

$$x_{SpT} = d_{Sp},\ x_{Sp0} = 0; \qquad p=1,\ldots,n_S \qquad (6.52)$$

$$x_{ijt} \geq 0, \text{integer}; \quad i=1,\ldots,S,\ j=1,\ldots,n_i, t=1,\ldots,T \qquad (6.53)$$

Constraint (6.47) shows that the necessary cumulative production of part type j of level i by the end of period t is determined explicitly by the quantity of final products assembled at level S, i.e., x_{ijt} depends on the final assembly schedule. Constraints (6.48) and (6.49) calculate the total cumulative production y_{it} of each level i during the first t periods. Constraint (6.50) guarantees that the total production of each final product during the first t periods is a nondecreasing function of t. Constraints (6.50), (6.51) and (6.53) ensure that exactly one product is scheduled for final assembly in every period. Finally, constraint (6.52) ensures that the total production requirements for each product are met.

The objective function (6.46) inherently captures the sequence dependent nature of the schedule for the lower level parts. The x_{ijt}, $i < S$ are calculated directly from the assembly schedule of the final products (the x_{Spt}), and the desired production goal for part type j of level i is calculated as the ideal proportion r_{ij} of the total cumulative production y_{it} of level i. Balanced schedules are created be keeping the production of all parts and final products as close to this goal as possible.

The min-max objective function (6.46) seeks a smooth schedule in every time period for every output. The value of (6.46) gives the maximum inventory or shortage with respect to the desired quantity of production that occurs at any time in the schedule. This information may be used for determining the number of kanbans or the necessary safety stocks.

6.4 Just-in-time scheduling of flexible assembly lines

Note that taking into account relations (6.47) and (6.49), the objective function (6.46) can be transformed into the following formula

$$\max_{i,j,t} \left| \left(\sum_{p=1}^{n_S} \gamma_{ijp} x_{Spt} \right) \right| \qquad (6.54)$$

where γ_{ijp} is a measure of the weighted deviations in the usage of part type j of level i from the proportional usage per unit of final product p.

$$\gamma_{ijp} = \lambda_{ij} \left(g_{ijp} - r_{ij} \sum_{k=1}^{n_i} g_{ikp} \right) \qquad (6.55)$$

This notation highlights the inherent pull nature of the JIT system. For each period, the deviation of actual cumulative production of any part of any level from desired cumulative production is determined explicitly by the final assembly schedule. The deviation calculated for part type j of level i depends solely on known a priori constants γ_{ijp} multiplied by the cumulative prodution x_{Spt} of the final level S.

In [59] a dynamic programming algorithm is developed for the min-max balanced scheduling problem (6.46) – (6.53).

Define $\Psi(x_{S1t}, \ldots, x_{Sn_St})$ to be the minimum value of the maximum deviation (6.54) for all parts and final products (all i and j) over all partial schedules which lead to state $(x_{S1t}, \ldots, x_{Sn_St})$ at the end of period t. The following dynamic programming recursion holds for $\Psi(x_{S1t}, \ldots, x_{Sn_St})$

$$\Psi(x_{S1t}, \ldots, x_{Sn_St}) = \min_{1 \leq p \leq n_S} \{\max\{\Psi(x_{S1t}, \ldots, x_{Spt} - 1, \ldots, x_{Sn_St}),$$

$$\max_{i,j}\{|(\sum_{p=1}^{n_S} \gamma_{ijp} x_{Spt})| : \sum_{p=1}^{n_S} x_{Spt} = t\}\} : x_{Spt} \geq 1\} \qquad (6.56)$$

$$\Psi(0, \ldots, 0) = 0 \qquad (6.57)$$

The first term in (6.56) is the maximum deviation (6.54) determined for all parts and final products over all partial schedules which lead to state $(x_{S1t}, \ldots, x_{Spt} - 1, \ldots, x_{Sn_St})$ for some $p = 1, \ldots, n_S$ at the end of period $t - 1$. The second term is the maximum deviation of actual production from desired production over all parts and final products in state $(x_{S1t}, \ldots, x_{Sn_St})$, at the end of period $t = \sum_{p=1}^{n_S} x_{Spt}$.

In [59] the time and space requirements of the dynamic programming algorithm have been analyzed and several techniques for improving its performance are described.

6.5 Multilevel scheduling of flexible assembly lines with limited intermediate buffers

In this section a hierarchical algorithm is proposed for scheduling lots of products in a multistage flexible assembly line with limited intermediate buffers. The approach is based on a multilevel programming formulation which implies a pull-type scheduling strategy typical for the just-in-time production systems.

Let us consider again the flexible assembly line with limited intermediate buffers described in Sect. 6.1. The line consists of $S \geq 2$ assembly stages in series with limited in-process buffers of capacity B_i between any two successive stages i and $i+1$ ($i = 1, \ldots, S-1$) where the in-process inventory can be maintained.

In the system N different products are assembled over a finite scheduling horizon. The horizon is made up of T assignment periods (shifts, for example), which have equal time-duration. During any period the assignment of products to machines is considered fixed and at most one product can be scheduled on each machine. Each product j ($j = 1, \ldots, N$) has to be assembled through stages $1, 2, \ldots, S$ in that order. As a result of processing j in stage i ($i = 1, \ldots, S-1$), the subasssembly A_{ij} is obtained, whereas the final assembly A_{Sj} is made in the final stage S. The subassembly A_{ij} produced in stage i ($i = 1, \ldots, S-1$) is transferred either directly to an available machine in the next stage $i+1$ or to a buffer ahead of that stage (see Fig. 6.10). Let α_{ij} be the inventory holding cost per unit of subassembly A_{ij} carried in the buffer between stages i and $i+1$ for one period. The transportation time required to transfer product from stage i to stage $i+1$ is denoted by q_i. Each product A_{i+1j} assembled in stage $i+1$ in period t is made up of subassembly A_{ij} produced in stage i in period $t - q_i$ at the latest.

Each assembly stage i ($i = 1, \ldots, S$) is made up of $M_i \geq 1$ identical parallel machines and each product j can be processed on any of M_i machines at a fixed production rate of r_{ij} units of A_{ij} per period. When product j is assigned to a machine in stage i, all r_{ij} units of A_{ij} must be made.

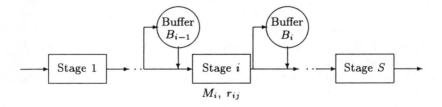

Fig. 6.10. A flexible assembly line with intermediate buffers

6.5 Multilevel scheduling of flexible assembly lines with limited buffers

Given is a production program $(d_1, \ldots, d_N)$ where d_j $(j = 1, \ldots, N)$ is the required number of machine-periods of production of the finished product A_{Sj}. The problem objective is to assign, in each stage, the products to the machines over the horizon in such a way as to complete the production program in a minimum time, with the in-process inventory holding cost at the lowest possible level.

The differences in production rates at different stages and for different products result in variations in the in-process inventories. Such variations should be limited. It is necessary that there are no shortages of the in-process inventories and that the buffer capacities are not exceeded. Moreover, the in-process inventory holding costs should usually be minimized. The in-process inventories maintained in the interstage buffers enable the successive stages to be operated relatively independently. The limited storage space of the buffers, however, may restrict the efficient utilization of the production capacities of all the stages.

The problem considered in this section is a lot size scheduling unlike the individual part scheduling problem presented in Sect. 6.1. Therefore a different approach is suggested here for its solution. The problem of lot size scheduling in a flexible assembly line will be formulated as a multilevel integer program, where each level problem is a single-stage scheduling problem. The interdependence of each production stage is modelled by jointly dependent constraints on the overall assembly schedule. Such constraints take into account the subassemblies availability and the interstage buffers limited capacity. An approximation hierarchical scheduling algorithm is provided and its possible application is illustrated with a numerical example.

6.5.1 Multilevel programming formulation

Define the problem decision variables u_{ijt} $(i = 1, \ldots, S, j = 1, \ldots, N, t = 1, \ldots, T)$ as the total number of machines assigned in stage i to produce product j in period t. The production schedule for stage i can be written as

$$u^i = \{u_{ijt} : j = 1, \ldots, N, t = 1, \ldots, T\} \tag{6.58}$$

In addition, the following auxiliary variables will be used in the sequel:

a_{ipt} $(i = 2, \ldots S, p = 1, \ldots, i-1, t = 1, \ldots, T)$ – the maximum number of machine-periods of production in stage p that can be processed in stage i $(i > p)$ during the first t periods

$$a_{ipt} = \max\{0, M_p(t - \sum_{k=1}^{i-1} q_k)\} \tag{6.59}$$

b_{it} $(i = 1, \ldots S-1, t = 1, \ldots, T)$ – the maximum production in stage i that can be carried in the buffer and/or used up in stage $i+1$ during the first t periods

154   6. Production Scheduling in Flexible Assembly Lines

$$b_{it} = B_i + \sum_{j=1}^{N}\sum_{k=1}^{t} r_{i+1j} u_{i+1jk} \qquad (6.60)$$

v_{ijt} ($i = 1,\ldots S-1$, $j = 1,\ldots,N$, $t = 1,\ldots,T$) – the minimum number of machine-periods of production of subassembly A_{ij} in stage i that must occur during the first t periods in order to meet the requirements for A_{ij}

$$v_{ijt} = \lceil (r_{i+1j}/r_{ij}) \sum_{k=1}^{t+q_i} u_{i+1jk} \rceil \qquad (6.61)$$

where $\lceil a \rceil$ denotes the smallest integer not less than a.

Local and interstage constraints. In general, each stage i of the assembly line can be characterized by an individual objective function f_i, $i = 1,\ldots,S$ defined over a jointly dependent schedule set U_0, which is to be minimized by the appropriate choice of schedule $u^i \in U_i$, where U_i is the set of schedules satisfying the local constraints in stage i.

The local constraints for stage i include:

— the machine assignment constraints

$$\sum_{j=1}^{N} u_{ijt} \le M_i; \quad t = 1,\ldots,T \qquad (6.62)$$

$$u_{ijt} \ge 0, \text{integer}; \quad j = 1,\ldots,N; \quad t = 1,\ldots,T \qquad (6.63)$$

— the input buffer capacity constraints

$$\sum_{j=1}^{N}(r_{i-1j}\lceil (r_{ij}/r_{i-1j})\sum_{k=1}^{t} u_{ijk}\rceil - r_{ij}\sum_{k=1}^{t-q_i-1} u_{ijk}) \le B_{i-1}; \; t = q_{i-1},\ldots,T \qquad (6.64)$$

where $B_0 = \infty$.
— the production requirement constraints

$$\sum_{t=1}^{T} u_{ijt} = d_{ij}; \quad j = 1,\ldots,N \qquad (6.65)$$

where d_{ij} is the required number of machine-periods of production of product j in stage i

$$d_{ij} = \lceil (r_{i+1j}/r_{ij})d_{i+1j} \rceil; \quad i = 1,\ldots,S-1; \quad j = 1,\ldots,N$$
$$d_{Sj} = d_j; \quad j = 1,\ldots,N$$

The set U_0 is defined by the following interstage constraints:

6.5 Multilevel scheduling of flexible assembly lines with limited buffers

— the output buffer capacity constraints

$$\sum_{j=1}^{N}\sum_{k=1}^{t} r_{ij} u_{ijk} \leq b_{it}; \; i=1,\ldots,S-1; \; t=1,\ldots,T \qquad (6.66)$$

— the subassemblies availability constraints

$$\sum_{j=1}^{N}\sum_{k=1}^{t} (r_{ij}/r_{pj}) u_{ijk} \leq a_{ipt}; \; p=1,\ldots,i-1; \; t=1,\ldots,T \qquad (6.67)$$

— the subassemblies requirement constraints

$$\sum_{k=1}^{t} u_{ijk} \geq v_{ijt}; \; i=1,\ldots,S-1; \; j=1,\ldots,N; \; t=1,\ldots,T \qquad (6.68)$$

The production schedules for the multilevel assembly line are determined sequentially, beginning with the schedule $u^S \in U_S$ for the final stage S, followed by schedule $u^{S-1} \in U_{S-1}$ for stage $S-1$, down through schedule $u^1 \in U_1$ for stage 1. The schedule determined for a later stage of the line affects the assignment of machines available for an earlier stage through U_0.

Now the multilevel programming formulation of the flexible assembly line scheduling problem can be written as follows [91, 93]:

Model MLS: *Multilevel scheduling of a flexible assembly line with limited intermediate buffers*

$$\begin{aligned}
& \min_{u^S \in U_S} f_S(u), \quad \text{where } u^{S-1} \text{ solves} \\
& \min_{u^{S-1} \in U_{S-1}} f_{S-1}(u), \quad \text{where } u^{S-2} \text{ solves} \\
& \qquad \vdots \\
& \min_{u^1 \in U_1} f_1(u) \\
& \text{subject to} \quad (u^1, \ldots, u^S) \in U_0
\end{aligned} \qquad (6.69)$$

The above formulation establishes the interdependence of each level problem. The problem of scheduling stage i is the minimization of function f_i over $U_0 \cap U_i$ for fixed values of $(u^{i+1}, \ldots, u^S)$. Thus the ith level problem can be interpreted as one parametrized in $(u^{i+1}, \ldots, u^S)$ and implicit in $(u^1, \ldots, u^{i-1})$.

Let us notice that the ith level problem has a feasible solution for a given schedule u^{i+1} if the following conditions are satisfied:

$$\sum_{j=1}^{N} v_{ijt} \leq M_i t; \ i = 1, \ldots, S-1, t = 1, \ldots, T \quad (6.70)$$

$$\sum_{j=1}^{N} r_{ij} v_{ijt} \leq b_{it}; \ i = 1, \ldots, S-1, t = 1, \ldots, T \quad (6.71)$$

$$\sum_{j=1}^{N} (r_{ij}/r_{pj}) v_{ijt} \leq a_{ipt}; \ i = 1, \ldots, S-1, \ p = 1, \ldots, i-1, t = 1, \ldots, T \quad (6.72)$$

The constraints (6.70), (6.71) and (6.72) are implied by the inequalities (6.62) and (6.68), (6.66) and (6.68), and (6.67) and (6.68), respectively. The condition (6.71) can be equivalently written as the following bound on the minimum capacity B_i of the stage i output buffer, given the schedule u^{i+1}

$$B_i \geq B_{imin} = \max_i \left\{ \sum_{j=1}^{N} r_{ij} \left[(r_{i+1j}/r_{ij}) \sum_{k=1}^{t+q_i} u_{i+1jk} \right] - \sum_{j=1}^{N} \sum_{k=1}^{t} r_{i+1j} u_{i+1jk} \right\} \quad (6.73)$$

In order to minimize the total cost of carrying in-process inventory between the stages and to minimize the makespan, the objective functions f_i ($i = 1, \ldots, S$) for various levels of (6.69) can be selected as below [91].

$$f_i(u) = -\sum_{j=1}^{N} \sum_{t=1}^{T} e_{ij} t u_{ijt}; \ i = 1, \ldots, S-1 \quad (6.74)$$

$$f_S(u) = C_{max} = T \quad (6.75)$$

where $e_{ij} = \alpha_{ij} r_{ij}$.

6.5.2 Algorithm for multilevel scheduling

The hierarchical scheduling algorithm based on the multilevel programming formulation constructs the assembly schedule using the pull strategy typical for the just-in-time production systems (see Fig.6.11).

The solution procedure consists of three scheduling algorithms, the first, $\mathcal{A}1$ – for scheduling the final stage S, the second, $\mathcal{A}2$ – for scheduling intermediate stages i, $i = 2, \ldots, S-1$, and the third, $\mathcal{A}3$ – for scheduling the first stage $i = 1$. The algorithms $\mathcal{A}1$ and $\mathcal{A}3$ are period-by-period heuristics, in which each assignment period is considered once. The algorithm $\mathcal{A}2$, however, is a recursive procedure capable of making adjustments in the schedules when a buffer is overfilled or shortage of a subassembly occurs (for a detailed description of the three scheduling algorithms, see [91]).

Period $t = 1$ is the first period assigned in the algorithm $\mathcal{A}1$, while the final period $t = T_i$ is assigned as the first one in the algorithms $\mathcal{A}2$ and $\mathcal{A}3$, where constraints (6.62), (6.64) and (6.66) imply the following formula:

6.5 Multilevel scheduling of flexible assembly lines with limited buffers

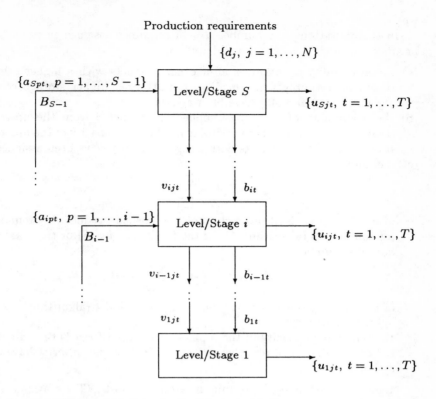

Fig. 6.11. Multilevel scheduling of a flexible assembly line

$$T_i = \max_{1 \leq t \leq T_{i+1} - q_i} \{t + \max\{0, \lceil (\sum_{j=1}^{N} r_{ij} d_{ij} - b_{it})/\gamma_i \rceil\}\} \quad (6.76)$$

where $\gamma_i = \max\{M_i r_{imax}, B_{i-1}/q_{i-1}\}$, $r_{imax} = \max_j\{r_{ij}\}$.

In the scheduling algorithm $\mathcal{A}1$ for the final stage S, the required lots of products with the longest total assembly times are scheduled first. The products are ordered according to nonincreasing assembly time required to complete the production requirements:

$$\beta_1 d_1 \geq \beta_2 d_2 \geq \cdots \geq \beta_N d_N \quad (6.77)$$

where $\beta_j = \max\{1/M_S, \max_{1 \leq p \leq S-1}\{r_{Sj}/r_{pj} M_p\}\}; \ j = 1, \ldots, N$.

In the scheduling algorithms $\mathcal{A}2$ and $\mathcal{A}3$ for stages $i = 1, 2, \ldots, S-1$, the required lots of products with the highest storage costs are scheduled last. The products are ordered according to nonincreasing holding costs beginning with $j = N$:

$$e_{iN} \geq e_{iN-1} \geq \cdots \geq e_{i1}, \ i = 1, \ldots, S-1 \quad (6.78)$$

where $e_{ij} = \alpha_{ij} r_{ij}$.

In the algorithms, the number u_{ijt} of machines assigned in period t to product j in stage i is determined based on:

(a) the availability of machines after other products with a higher priority (a longer assembly time in algorithm $\mathcal{A}1$ or a higher storage cost in algorithms $\mathcal{A}2$ and $\mathcal{A}3$) have been assigned;
(b) the availability of the subassemblies of product j from the upstream stages $1, \ldots, i-1$, i.e., the remaining production capacity of the upstream stages after other products with a higher priority have been assigned;
(c) the number

$$(d_j - \sum_{k=1}^{t-1} u_{Sjk})$$

of machine-periods of production of the finished product A_{Sj} remaining for completion (algorithm $\mathcal{A}1$ for the final stage $i = S$) or the maximum desirable number

$$(d_{ij} - v_{ijt-1} - \sum_{k=t+1}^{T_i} u_{ijk})$$

of machine-periods of production of A_{ij} in period t (algorithms $\mathcal{A}2$ and $\mathcal{A}3$ for stages $i < S$);
(d) the remaining capacity of the input buffer B_{i-1} (if $i > 1$) or the output buffer B_i (if $i < S$) after other products with a higher priority have been assigned.

Algorithms $\mathcal{A}1$ and $\mathcal{A}3$ terminate after at most NT iterations, since each product is considered at most once during at most T periods. Algorithm $\mathcal{A}2$, however, may require NT^2 iterations owing to the corrective steps which result in recursive determination of schedule for stages $i = 2, \ldots, S - 1$. Therefore, computational complexity of the multilevel algorithm is $O(SN \log N + SNT^2)$ [91, 93].

It is worth noting that algorithm $\mathcal{A}3$ gives an optimal solution to the first-level problem if one machine-period of production of each product A_{1j} ($j = 1, \ldots, N$) takes up the same amount of space of the buffer B_1 (cf. [91]). In such a case B_1 can be redefined as the maximum number of machine-periods of production in stage 1 that can be carried in inventory at any time, and $r_{1j} = 1$ for all j.

Finally, the following lower bounds can be easily derived on the optimal values of the objective functions f_i, $i = 1, \ldots S$ (6.74) and (6.75), see [91]

$$LB_i = \sum_{j=1}^{N} e_{ij} (\sum_{t=1}^{T_i-1} v_{ijt} - T_i d_{ij}); \; i = 1, \ldots, S - 1 \qquad (6.79)$$

$$LB_S = \max_{1 \le i \le S} \{\lceil \sum_{j=1}^{N} d_{ij}/M_i \rceil, \max_{1 \le p \le i-1} \lceil \sum_{j=1}^{N} (r_{ij}/r_{pj}) d_{ij}/M_p + \sum_{k=1}^{i-1} q_k \rceil\} \qquad (6.80)$$

6.5.3 Numerical example

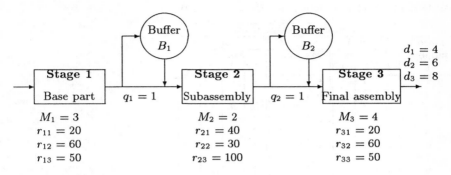

Fig. 6.12. A three-stage assembly line

In this subsection, an illustrative example for a three-stage production line is presented (see Fig. 6.12), where three product types ($N = 3$) are assembled [91]. The line is made up of $M_1 = 3$ machines in stage 1 for base part processing, $M_2 = 2$ machines in stage 2 for subassembly and $M_3 = 4$ machines for final assembly. The corresponding production rates r_{ij} ($i = 1, 2, 3$, $j = 1, 2, 3$) are shown in Fig. 6.12 and the production requirements for the final products are:
$d_1 = 4$ machine-periods of production of product type 1,
$d_2 = 6$ machine-periods of production of product type 2,
$d_3 = 8$ machine-periods of production of product type 3.

The interstage buffer capacities are $B_1 = B_2 = 100$ and the transportation times between the stages are $q_1 = q_2 = 1$. The base parts and subassemblies cause similar holding costs, that is, $\alpha_{ij} = 1$ ($i = 1, 2$, $j = 1, 2, 3$). Therefore, the minimization of the total amount of in-process inventories is the objective for the first two stages. The scheduling horizon is made up of $T = 15$ periods.

The production schedule obtained for the example using the multilevel scheduling algorithm is presented in Table 6.11 where for each period the product numbers assigned to machines are indicated.

The values of the objective functions are:

$$f_1(u^1) = -4640, \quad f_2(u^2) = -5660, \quad f_3(u^3) = C_{max} = 14$$

where the lower bounds (6.79) and (6.80) are

$$LB_1 = -4640, \quad LB_2 = -5840, \quad LB_3 = 11.$$

If the buffer capacities were not fixed and could be extended to

$$B_1 = q_1 \max\{M_1 r_{1max}, M_2 r_{2max}\} = 200, \quad B_2 = q_2 M_2 r_{2max} = 200,$$

Table 6.11. Production schedule ($B_1 = B_2 = 100$)

Period number t	Machine assignments								
	Stage 1 Machine			Stage 2 Machine		Stage 3 Machine			
	1	2	3	1	2	1	2	3	4
1	3	3	-	-	-	-	-	-	-
2	3	3	-	3	-	-	-	-	-
3	3	3	-	3	-	3	3	-	-
4	3	3	-	3	-	3	3	-	-
5	2	-	-	3	-	3	3	-	-
6	2	-	-	2	-	3	3	-	-
7	2	-	-	2	2	-	-	-	-
8	2	-	-	2	2	2	-	-	-
9	2	-	-	2	2	2	-	-	-
10	1	1	-	2	2	2	-	-	-
11	2	-	-	2	1	2	-	-	-
12	1	1	-	2	-	2	1	1	-
13	-	-	-	2	1	-	-	-	-
14	-	-	-	-	-	2	1	1	-
Legend: (-) idle machine									

Table 6.12. Production schedule ($B_1 = B_2 = 200$)

Period number t	Machine assignments								
	Stage 1 Machine			Stage 2 Machine		Stage 3 Machine			
	1	2	3	1	2	1	2	3	4
1	3	3	3	-	-	-	-	-	-
2	3	3	3	3	-	-	-	-	-
3	2	-	-	3	3	3	3	-	-
4	3	3	-	2	-	3	3	3	3
5	2	-	-	2	3	-	-	-	-
6	2	-	-	2	2	2	3	3	-
7	2	-	-	2	2	2	-	-	-
8	2	-	-	2	2	2	-	-	-
9	2	1	-	2	2	2	-	-	-
10	1	1	1	2	2	2	-	-	-
11	-	-	-	1	1	2	-	-	-
12	-	-	-	-	-	1	1	1	1
Legend: (-) idle machine									

6.5 Multilevel scheduling of flexible assembly lines with limited buffers

then the completion time would be the shortest possible and the algorithms yield the schedule shown in Table 6.12. In this case

$$f_1(u^1) = -3910, \ f_2(u^2) = -4850, \ f_3(u^3) = C_{max} = 12$$

and the corresponding lower bounds (6.79) and (6.80) are

$$LB_1 = -3980, \ LB_2 = -4880, \ LB_3 = 12.$$

The hierarchical approach proposed enables lot-sizing and scheduling to be integrated for the multistage flexible assembly line within the multilevel programming framework. The scheduling heuristics prove quite satisfactory both in the worst-case and in an average performance [91, 92].

7. Machine and Vehicle Scheduling in Flexible Assembly Systems

A *flexible assembly system* can be considered to be made up of two interrelated subsystems: an assembly subsystem and a materials handling subsystem. The two subsystems are closely integrated so that the performance of one affects the other. While completion of each assembly operation at an assembly station generates an arrival of some transportation task to the AGV subsystem, completion of a transportation task by an AGV determines new assembly task for some assembly station.

The FAS scheduling is made in a dynamic scheduling environment. The system status changes so frequently that at one time the assembly subsystem can become a "critical resource" whereas at some other time the AGV subsystem becomes a critical resource and dominates the schedule. Such a two-way interaction requires that in FAS scheduling the two subsystems must be taken into account simultaneously. Therefore, machine and vehicle schedules in a FAS should be determined simultaneously and if possible on-line and close to real time, e.g., [87, 88].

On-line scheduling attempts to schedule operations one at a time when they are needed whereas *off-line scheduling* refers to scheduling all operations of available products for the entire scheduling horizon. On the other hand, *real-time scheduling* is a short-term decision making process which generates and updates the schedule based on the current status of the system and the overall system requirements. Real-time scheduling can be made by either an off-line or an on-line method or a combination of methods. If off-line scheduling methods are used, the scheduling process becomes scheduling and rescheduling whereas for the on-line scheduling approach, the scheduling decision is made when the state of the system changes, such as task completion, vehicle arrival, etc. There are various advantages and disadvantages of each approach. Off-line scheduling requires large computational effort to frequently generate and update the schedule in a dynamic environment. On the other hand, scheduling decisions made by on-line methods may not provide the best results due to lack of a broad system view.

An important feature of the FAS scheduling is that both the input and output buffers at stations are limited. Therefore, there is always a possibility that a particular station can be blocked or the system can be locked due to limited buffer spaces. Blocking occurs when a station cannot move its product to a buffer if the buffer is full, whereas locking occurs when the system is totally prevented from functioning, i.e., no product movement can be achieved in the system. The limited buffer capacities must be taken into account in scheduling of a FAS.

The objective of the detailed machine and vehicle scheduling is to find an assignment of assembly operations to machines for each individual unit of each product type over a scheduling horizon as well as the associated time table for vehicle movements so as to complete a production order and minimize some optimality criterion.

A typical optimality criterion of the machine and vehicle schedule is to minimize its makespan, mean flow time or maximum lateness for a given set of product due dates, e.g., [17, 85, 87, 88, 96, 125].

There are two main approaches to be used for simultaneous machine and vehicle scheduling in a FAS, e.g., [99, 100]:

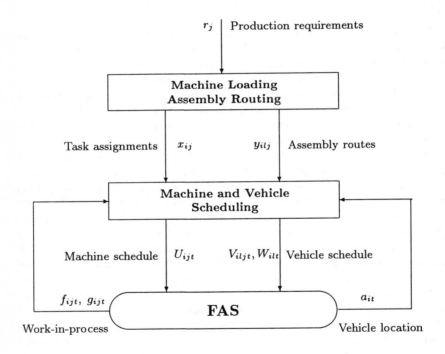

Fig. 7.1. Multi-level machine and vehicle scheduling in a FAS

1. *Multi-level approach* (see Fig. 7.1), in which first machine loading and assembly routing problems are solved and then, given task assignments

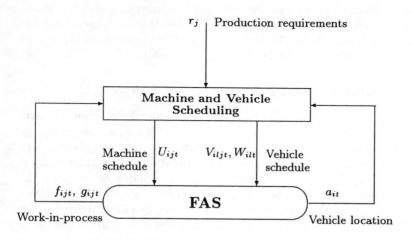

Fig. 7.2. Single-level machine and vehicle scheduling in a FAS

and assembly routes selected, detailed machine and vehicle schedules are determined for each individual unit of each product type.
2. *Single-level approach* (see Fig. 7.2), in which detailed assembly schedules for each individual unit of each product type and the corresponding time tables for vehicle movements are directly determined with no initial loading and routing decisions required.

The above two approaches will be discussed throughout this chapter.

7.1 Dispatching scheduling

Dispatching algorithms are widely used for scheduling in industrial practice, e.g., [45, 75, 84, 88]. The algorithms are based on various dispatching rules which prioritize the products for assignment to machines and vehicles. In general, the rules act as an independent local scheduling mechanism. Machine scheduling rules do not consider the availability of AGVs when the priorities of products are set. Similarly, AGV scheduling rules do not directly account for availability of machines for products to be assembled.

The procedures for scheduling of assembly operations on machines and scheduling of transportation operations on vehicles in a FAS are closely interrelated, e.g., [87]. The dispatching algorithms for machine and vehicle scheduling consider the various interactions between machines and AGVs. using information on assembly process and the system status. Some of this information is product related such as processing and transportation times, precedence relations among operations, product assembly routes, etc. Others are system related and depend on the status of machines and AGVs, planned and actual machine workloads, planned and actual transfer of products between machines, queue levels, the number of products in the system, current location of AGVs, etc., (see [32, 36, 67, 75, 86, 87, 88]).

7.1.1 Dispatching scheduling of assembly operations

The procedure for scheduling assembly operations on the machines is executed whenever a station completes performing of its current assembly operation and becomes available for the next assembly task assignment. For each candidate product waiting for its next operation in the station input buffer, the priority index is calculated. The priority index depends on the optimality criterion selected. For example, for makespan criterion, the index is a function of the remaining processing and transportation times, and for a due date related criterion, modified operation due date can be used as a substitute of the remaining processing and transportation times.

After the priority index is calculated for each operation of each product, the product waiting for its next operation with the highest priority index is loaded on the machine first.

7.1.2 Dispatching scheduling of transportation operations

The procedure for scheduling of transportation operations on the vehicles is executed whenever an AGV completes its current operation and becomes available for the next transportation task assignment.

In general, the procedure can have the following four hierarchical levels suggested in [87].

Level 1: **Checking the critical stations.**

Critical station is searched for, which is either blocked or its input and output buffers are full. Such stations neither can accept any product from the other stations nor can perform the operations because of blocking. Therefore, one of the outgoing products at this station has to be delivered to its next station. This decision can be made based on the following set of hierarchical rules:

1. A finished product at a critical station waiting to be transferred to the unload station is transferred first.
2. A product at the most demanded critical station is serviced first.
3. The highest priority is given to the product which has the smallest input queue level at the next station.
4. The highest priority is given to the product which is closest to the current location of idle AGV.
5. The highest priority is given to the product with the least/most amount of work remaining or the earliest due-date, depending on the scheduling criterion employed.

Level 2: **Checking the idle stations.**
If there are some idle stations, then other station is searched for to locate a station which can immediately deliver a product to this idle station. In the case where there is more than one idle station and more than one station can deliver the products to one of these stations, the following decision criteria can be employed to schedule the AGV for the next trip:
1. The highest priority is given to the station which is nearest to the current location of the idle AGV.
2. The highest priority is given to the product with the least/most amount of work remaining or the earliest due-date, depending on the scheduling criterion employed.

Level 3: **Checking the products in the central buffer.**
If the system has a central buffer and there are products in the buffer, a product with the most available destination input buffer space is transferred first. In the case of a tie, a product with the most amount of work remaining or the earliest due-date is selected, depending on the scheduling criterion employed.

Level 4: **Selecting the product for transfer and the destination station.**
If the central buffer is empty and there are no critical or idle stations then the system can be considered to be stable. In such a case, a product which has the highest chance of being processed earliest at its next station is selected to be transferred by an AGV first.

7.2 Machine and vehicle scheduling – a multi-level approach

Consider a FAS composed of m assembly stations $i \in I = \{1, \ldots, m\}$ of various types, loading station L, unloading station U, and a AGVs that permit products to move between any pair of stations. There are two kinds of trips performed by vehicles: loaded trips and deadheading trips. In a loaded trip, a product is taken from the station where an assembly task has been completed to the station where its next assembly task is assigned. Upon completing a loaded trip the vehicle either begins its next loaded trip if a product has been waiting for transfer at the vehicle last destination station or travels empty to its next pick-up station otherwise. The latter type of trips are called deadheading trips. Their durations are sequence-dependent and are not known until the vehicle schedule is specified. An appropriate selection of the deadheading trips has an effect on the quality of entire machine and vehicle schedule.

The set of stations is capable of performing n different operations required to assemble v different product types. Let $J = \{1, \ldots, n\}$ be the set of all individual operations of all product types, where each operation $j \in J$

Table 7.1. Notation

		Indices
i	=	machine i, $i = 1, \ldots, m$
j	=	operation j, $j = 1, \ldots, n$
t	=	assignment period t, $t = 1, \ldots, H$
		Input parameters
a	=	number of vehicles
b_i	=	capacity of buffer at machine i
p_{ij}	=	assembly time of operation j on machine i
q_{il}	=	transportation time of a vehicle from machine i to machine l; $i, l \in I$
r_j	=	production requirement for operation j (number of units of the product type that requires operation j)
x_{ij}	=	number of operations j assigned to machine i
y_{ilj}	=	number of products to be transferred after completion of operation j on machine i to machine l to perform next operation $j + 1$
		Decision Variables
U_{ijt}	=	1, if operation j is assigned to machine i in period t; otherwise $U_{ijt} = 0$
V_{iljt}	=	1, if transferring of product from machine i after completing operation j to machine l to perform operation $j+1$ starts at the beginning of period t; otherwise $V_{iljt} = 0$
W_{ilt}	=	1, if at the beginning of period t an empty vehicle starts moving from machine i to machine l to pick up a product waiting there for transfer; otherwise $W_{ilt} = 0$
T_t	=	length of assignment period t
		Auxiliary Variables
a_{it}	=	number of idle vehicles waiting at machine i at the beginning of period $t+1$
c_{ijt}	=	cumulative processing time of operation j on machine i during the first t periods
f_{ijt}	=	the completed fraction of operation j performed on machine i by the end of period t, $0 \le f_{ijt} < 1$
g_{ijt}	=	the inventory of operations j waiting in buffer at machine i at the beginning of period $t+1$
C_j	=	completion time of all r_j operations j

corresponds to only one product type. The assemble of each product type k ($k = 1, \ldots, v$) requires a sequence $J_k = [j_{k-1} + 1, j_{k-1} + 2, \ldots, j_k - 1, j_k]$ ($j_0 = 0$) of successively numbered operations to be performed in the listed order. The assembly process of each product type k begins with a loading operation $j_{k-1} + 1 \in J_L$ at loading station L and ends with an unloading operation $j_k \in J_U$ at unloading station U, where $J_L \subset J$ and $J_U \subset J$ are the subsets of loading and unloading operations for all product types, respectively. Each assembly operation $j \in J \setminus (J_L \bigcup J_U)$ can be performed at any station in the subset $I_j \subset I$ of assembly stations capable of performing j.

The base part of each product to be assembled is mounted on a pallet. Let P be the maximum number of pallets to be allowed in the system simultaneously. In process inventory can be held at each station i in finite buffer with

7.2 Machine and vehicle scheduling – a multi-level approach   169

storage space for b_i pallets with products. Finally, let r_j be the production requirement for operation j (in number of units of a particular product type).

The problem objective is to determine the machine and vehicle assignment over a scheduling horizon so as to meet all product requirements in a minimum time.

The basic notation used to describe the machine and vehicle scheduling problem is given in Table 7.1. The problem is formulated under the following *basic assumptions*:

- No assembly or transportation operation can be preempted.
- Each machine can process and each vehicle can transport at most one product at a time.
- The scheduling horizon is made up of H assignment periods (H is an unknown integer) which may have unequal time durations (see Fig. 7.3). During every period the assignment of operations to machines and vehicles to trips (pairs of machines) is considered fixed.
- During the scheduling horizon each machine must complete the assigned assembly tasks x_{ij}, and the vehicles must complete the associated transportation tasks y_{ilj} determined at the machine loading/assembly routing level.
- Transportation times of loaded and unloaded (empty) vehicles are equal and the vehicles move with no delay due to congestion (see Fig. 7.4).
- Each machine (including loading/unloading station) is supplied with a finite input/output buffer in which products can wait for their next operations on this machine or for transfer to other machines.
- Each pallet can carry only one product at a time and at most P pallets are allowed in the system simultaneously.
- At the beginning all vehicles wait in the central depot, where the central depot, loading and unloading stations are located at one place.

7.2.1 Problem variables

The following four sets of *decision variables* are introduced to model the machine and vehicle scheduling problem (see Fig. 7.3, 7.4).

- Machine assignment variables $U_{ijt} \in \{0,1\}$, where $U_{ijt} = 1$ if operation j is assigned to machine i in period t; otherwise $U_{ijt} = 0$.
- Loaded vehicle assignment variables $V_{iljt} \in \{0,1\}$, where $V_{iljt} = 1$ if transferring of product from machine i to machine l to perform operation $j+1$ starts at the beginning of period t; otherwise $V_{iljt} = 0$.
- Idle (empty) vehicle assignment variables $W_{ilt} \in \{0,1\}$, where $W_{ilt} = 1$ if at the beginning of period t an empty vehicle starts moving from machine i to machine l to pick up a product waiting there for transfer between machines; otherwise $W_{ilt} = 0$.
- Length $T_t > 0$ of assignment period t.

170   7. Machine and Vehicle Scheduling in Flexible Assembly Systems

In the definitions above the beginning of period t is the sum of the durations of all previous periods, i.e., $\sum_{s=1}^{t-1} T_s$.

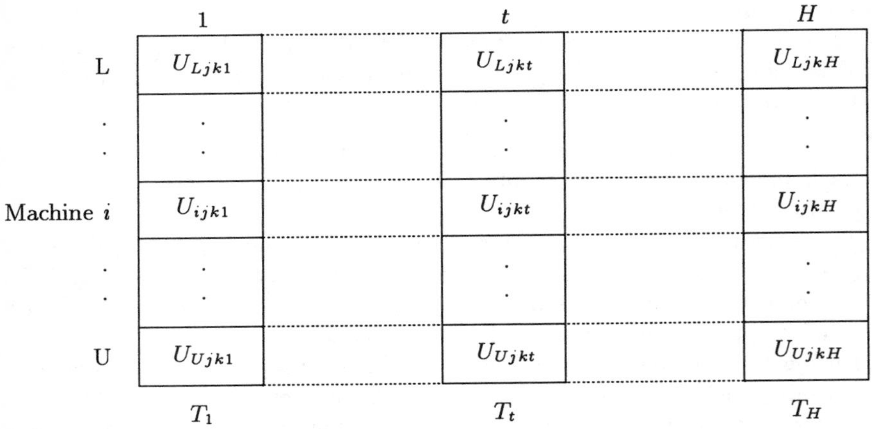

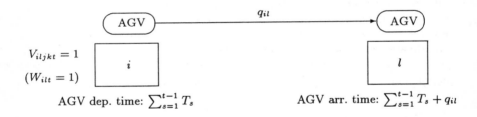

Fig. 7.3. Machine assignment variables

Fig. 7.4. Vehicle assignment variables

In addition, the following *auxiliary variables* will also be used in the sequel:

c_{ijt} $(i \in I_j,\ j \in J,\ t = 1,\ldots,H)$ – cumulative processing time of operations j on machine i during the first t periods

$$c_{ijt} = \sum_{s=1}^{t} T_s U_{ijs} \qquad (7.1)$$

f_{ijt} $(i \in I_j,\ j \in J,\ t = 1,\ldots,H)$ – the completed fraction of operation j performed on machine i by the end of period t, $0 \leq f_{ijt} < 1$

$$f_{ijt} = c_{ijt}/p_{ij} - \lfloor c_{ijt}/p_{ij} \rfloor \qquad (7.2)$$

g_{ijt} ($i \in I$, $j \in J$, $t = 1, \ldots, H$) – the inventory (number of products of a particular type) of operations j waiting in buffer at machine i at the beginning of period $t+1$

$$g_{ijt} = g_{ij0} + \lfloor c_{ijt}/p_{ij} \rfloor - \lceil c_{ij+1t}/p_{ij+1} \rceil + \sum_{l \neq i}(\sum_{s=1}^{e_{lit}} V_{lijs} - \sum_{s=1}^{t} V_{iljs}) \quad (7.3)$$

where
$\lfloor c \rfloor$, $\lceil c \rceil$ is, respectively, the greatest integer not greater than c, the smallest integer not less than c,
g_{ij0} is the beginning in-process inventory,
e_{lit} is the latest starting period for a vehicle transferring a product from machine l to machine i such that it can reach the destination machine by the end of period t

$$e_{lit} = \max\{r : \sum_{s=r}^{t} T_s \geq q_{li}\}$$

a_{it} ($i = 0, \ldots, m$, $t = 1, \ldots, H$) – number of idle vehicles waiting at machine i at the beginning of period $t+1$, i.e. at time $\sum_{s=1}^{t} T_s$

$$a_{0t} = a + \sum_{l=1}^{m}[\sum_{s=1}^{e_{l0t}}(W_{l0s} + \sum_{j \in J} V_{lm+1js}) - \sum_{s=1}^{t}(W_{0ls} + \sum_{j \in J_L} V_{0ljs})] (7.4)$$

$$a_{it} = \sum_{l \neq i}[\sum_{s=1}^{e_{lit}}(W_{lis} + \sum_{j \in J} V_{lijs}) - \sum_{s=1}^{t}(W_{ils} + \sum_{j \in J} V_{iljs})]; \quad i = 1, \ldots, m (7.5)$$

where $i = 0$ denotes central depot at which all a vehicles wait at the beginning, which is located at one place with loading and unloading stations.

T_t^p ($t = 1, \ldots, H$) – the shortest remaining processing time of the operations performed in period t

$$T_t^p = \min_{\{(i,j):U_{ijt}=1\}} \{p_{ij}(1 - f_{ijt-1})\} \quad (7.6)$$

T_t^q ($t = 1, \ldots, H$) – the shortest remaining transportation time of the vehicles moving in period t

$$T_t^q = \min_{\{(i,l):V_{iljr}=1 \text{ or } W_{ilr}=1, r \leq t\}} \{q_{il} - \sum_{s=r}^{t-1} T_s : q_{il} > \sum_{s=r}^{t-1} T_s\} \quad (7.7)$$

7.2.2 Constraints

The algorithm for scheduling the release of products into the system, finished products unloading from the system and scheduling of machines and vehicles is called **SMAV-ML** (Scheduling of Machines And Vehicles – Multi-Level) [105, 108]. It is a hierarchical period-by-period heuristic in which each assignment period is considered once. The decision variables for each assignment period are determined in the following sequence, see Fig. 7.5 (the lowercase letters are used below for the decision variables that have already been determined at an earlier stage of the decision process):

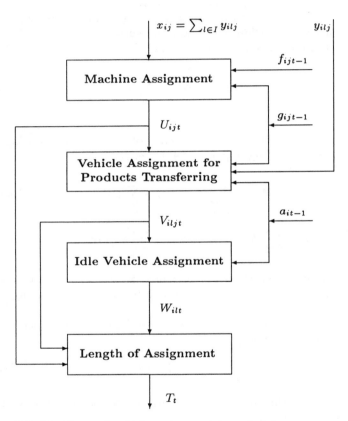

Fig. 7.5. Machine and vehicle assignment in period t

1. **Machine assignment** (U_{ijt}; $i \in I_j$; $j \in J$)
   subject to

$$f_{ijt-1} + u_{ijt-1} - 1 \leq U_{ijt} \leq g_{ij-1t-1} + \lceil f_{ijt-1} \rceil; \quad i \in I_j; \quad j \in J \quad (7.8)$$

$$U_{ijt} \leq x_{ij} - \lfloor c_{ijt}/p_{ij} \rfloor; \quad i \in I_j; \quad j \in J \quad (7.9)$$

$$\sum_{j \in J} U_{ijt} \leq 1; \quad U_{ijt} \in \{0,1\}; \quad i \in I_j; \quad j \in J \quad (7.10)$$

Constraints (7.8) insure against preemption of uncompleted operations and account for the inventory of products available for assignment at each machine. Constraint (7.9) requires that operation assignments determined at the machine loading level be met. Constraints (7.10) ensure that at most one operation is assigned to each machine at a time.

2. **Products loading assignment** $(U_{0jt}; \ j \in J_L)$
subject to

$$\sum_{j \in J_L} U_{0jt} \leq P - \sum_{i=0}^{m} \sum_{j \in J} g_{ijt-1} - \sum_{i \in I} \sum_{j \in J} \lceil f_{ijt-1} \rceil \quad (7.11)$$

$$\sum_{j \in J_L} U_{0jt} \leq 1; \quad U_{0jt} \in \{0,1\}; \quad j \in J_L \quad (7.12)$$

Constraint (7.11) prevents loading of a new product unless the total number of products in the system is less than P, the maximum number of pallets allowed simultaneously. Constraints (7.12) prevent loading of more than one product at a time.

3. **Vehicle assignment for products transferring to the destination machines** $(V_{iljt}; \ i \neq m+1; \ l \neq 0; \ i \neq l)$
subject to

$$\sum_{l \in I} V_{iljt} \leq g_{ijt-1} + \lceil f_{ij+1t-1} \rceil - u_{ij+1t}; \quad i \in I_j; \quad j \in J \setminus J_L \quad (7.13)$$

$$\sum_{j \in J} \sum_{l \neq i} V_{iljt} \leq a_{it-1}; \quad i \in I \quad (7.14)$$

$$V_{iljt} \leq y_{ilj} - \sum_{s=1}^{t-1} v_{iljs}; \quad i,l \in I; \quad i \neq l; \quad j \in J \quad (7.15)$$

$$\sum_{i \in I} \sum_{j \in J} V_{iljt} \leq b_l - \sum_{j \in J}(g_{ljt-1} + u_{ljt-1} - u_{ljt}); \quad l \in I \quad (7.16)$$

$$V_{iljt} \in \{0,1\}; \quad i,l \in I; \quad i \neq l; \quad j \in J \quad (7.17)$$

Constraint (7.13) accounts for the inventory of products waiting for transfer and (7.14) accounts for location of idle vehicles. Constraint (7.15) requires that intermachine flows of products determined at the assembly routing level be met. Constraint (7.16) prevents the buffer from overfilling.

4. **Idle vehicle assignment to the sending machines** $(W_{ilt}; \ i \neq l; \ l \neq m+1)$
subject to

$$\sum_{l \neq i} W_{ilt} \leq a_{it-1} - \sum_{j \in J} \sum_{l \neq i} v_{iljt}; \quad i \in I \tag{7.18}$$

$$W_{ilt} \leq \sum_{j \in J} \sum_{k \neq l} (y_{lkj} - \sum_{s=1}^{t} v_{lkjs}); \quad i, l \in I; \quad i \neq l \tag{7.19}$$

$$W_{ilt} \leq \sum_{j \in J} (g_{ljt-1} - \sum_{k \neq l} v_{lkjt}); \quad i, l \in I; \quad i \neq l \tag{7.20}$$

$$W_{ilt} \in \{0, 1\}; \quad i, l \in I; \quad i \neq l \tag{7.21}$$

Constraint (7.18) accounts for location of the remaining idle vehicles. Constraints (7.19) and (7.20) ensure the assignment of idle vehicles to the machines with the inventory of products required transferring.

5. **Length of assignment period** (T_t)

$$T_t = \min\{T_t^p, T_t^q\} \tag{7.22}$$

7.2.3 Dispatching rules for machine and vehicle scheduling

In every period the assignment decisions are made using several dispatching rules. A priority rule is used to select the next operation for assignment to an idle machine from the set of products waiting in its buffer. Similarly, priority rules are used to select products for transfer to other machines as well as to select the next movement of each idle vehicle (see [99, 100, 105, 108]).

In order to describe different priority rules used in the scheduling algorithm the following *auxiliary parameters* are introduced below:

P_j – total processing time of all operations j

$$P_j = \sum_{i \in I_j} p_{ij} x_{ij} \tag{7.23}$$

P_{jt} – processing time of operations j remaining at the beginning of period t

$$P_{jt} = P_j - \sum_{i \in I_j} c_{ijt-1} \tag{7.24}$$

Q_j – total transportation time required to transfer all products with completed operation j and waiting for next operation $j+1$

$$Q_j = \sum_{i \in I_j} \sum_{l \in I_{j+1}} q_{il} y_{ilj} \tag{7.25}$$

Q_{jt} – total transportation time Q_j calculated at the beginning of period t

$$Q_{jt} = Q_j - \sum_{i \in I_j} \sum_{l \in I_{j+1}} q_{il} \sum_{s=1}^{t-1} V_{iljs} \tag{7.26}$$

R_j – total work remaining (processing and transportation time) required to complete all remaining operations beginning with j and ending with unloading of all products of a particular type

$$R_j = P_j + Q_j + R_{j+1}; \quad j \in J \setminus J_U \tag{7.27}$$

$$R_j = P_j; \quad j \in J_U \tag{7.28}$$

R_{jt} – total work remaining R_j calculated at the beginning of period t

$$R_{jt} = P_{jt} + Q_{jt} + R_{j+1t}; \quad j \in J \setminus J_U \tag{7.29}$$

$$R_{jt} = P_{jt}; \quad j \in J_U \tag{7.30}$$

Dispatching rules for machine assignment.
The following priority rules are used to determine the assignment U_{ijt} of operations to machines for products waiting in their buffers. The operation j with the minimum (or maximum) value of the priority index shown below is assigned first (subject to constraints (7.8)–(7.12)).

STP&TT (Shortest Total Processing & Transportation Time):
$\min_j \{P_j + Q_j\}$
SUP&TT (Shortest Uncompleted Processing & Transportation Time):
$\min_j \{P_{jt} + Q_{jt}\}$
LWKR (Least Work Remaining): $\min_j \{R_j\}$
LUWKR (Least Uncompleted Work Remaining): $\min_j \{R_{jt}\}$
MWKR (Most Work Remaining): $\max_j \{R_j\}$
MUWKR (Most Uncompleted Work Remaining): $\max_j \{R_{jt}\}$

Dispatching rules for vehicle assignment for product transferring.
The priority rules used to determine the assignment V_{iljt} of products (operations) to idle vehicles for transfer between machines are similar to those for machine assignment. The product with completed operation j waiting in buffer of machine i for transfer to another machine to perform its next operation $j+1$ with the minimum (or maximum) value of the priority index shown below is transferred first (subject to constraints (7.13)–(7.17)).

STP&TT (Shortest Total Processing & Transportation Time):
$\min_j \{Q_j + P_{j+1}\}$
SUP&TT (Shortest Uncompleted Processing & Transportation Time):
$\min_j \{Q_{jt} + P_{j+1t}\}$
LWKR (Least Work Remaining): $\min_j \{Q_j + R_{j+1}\}$
LUWKR (Least Uncompleted Work Remaining): $\min_j \{Q_{jt} + R_{j+1t}\}$
MWKR (Most Work Remaining): $\max_j \{Q_j + R_{j+1}\}$
MUWKR (Most Uncompleted Work Remaining): $\max_j \{Q_{jt} + R_{j+1t}\}$

Dispatching rules for machine selection
Having assigned the product with completed operation j to an idle vehicle for transfer from machine i, the machine l with the minimum (or maximum) value of the priority index shown below is selected as its next destination (subject to constraints (7.15), (7.16)).

STT (Shortest Transportation Time): $\min_l\{q_{il}\}$
SQS (Smallest Queue Size): $\min_l\{\sum_{j\in J} g_{ljt-1}\}$
LQS (Largest Queue Size): $\max_l\{\sum_{j\in J} g_{ljt-1}\}$
LW&TT (Largest Workload & Transfer Time): $\max_l\{\sum_{j\in J}(p_{lj}x_{lj}+q_{il}y_{ilj})\}$
LUW&TT (Largest Uncompleted Workload & Transfer Time):
$$\max_l\{\sum_{j\in J}(p_{lj}x_{lj} - \sum_{s=1}^{t-1} T_s U_{ljs}) + \sum_{j\in J} q_{il}(y_{ilj} - \sum_{s=1}^{t-1} V_{iljs})\}$$

Dispatching rules for idle vehicle assignment

The assignment W_{ilt} of remaining idle vehicles to their next machines is made using the dispatching rules proposed above for vehicle assignment for product transferring (subject to constraints (7.18)–(7.21)). An idle vehicle waiting at machine i is send to such a machine l at which a product is waiting for transfer to perform its next operation $j+1$ with the highest priority.

7.2.4 Scheduling algorithm

In *STEP 0* of the SMAV-ML algorithm presented below one should choose priority index for machine assignment U_{ijt} (including the loading and unloading stations), for vehicle assignment for product transferring V_{iljt}, and for idle vehicle assignment W_{ilt}.
In addition, priority indices are chosen to select the destination machines l for vehicles transferring products (variables V_{iljt}).

Algorithm SMAV-ML

STEP 0. Selection of the priority indices and the initial conditions.
   Select priority index for:
   – machine assignment,
   – vehicle assignment for product transferring,
   – selection of destination machines for loaded vehicles,
   – selection of destination machines for empty vehicles.
   Set:
   $a_{00} = a$, $a_{i0} = 0$, $i \in I$
   $c_{ij0} = 0$, $f_{ij0} = 0$, $g_{ij0} = 0$, $i \in I_j$, $j \in J$
   $t = 1$

STEP 1. Machine and products loading/unloading assignment in period t
   The highest priority operations j for the remaining tasks x_{ij} assign to idle machines i at which the products are waiting.

STEP 2. Vehicle assignment for products transferring in period t
   The highest priority operations j in the remaining transportation tasks y_{ilj} assign to idle vehicles, and then send to the destination machines with the highest priority at which empty buffers are available.

STEP 3. Idle vehicle assignment in period t
   An idle vehicle at machine i send to such a machine l where the product is waiting for transfer to perform operation $j+1$ with the highest priority.

STEP 4. Length of assignment period t

Set T_t equal to the longest possible duration (7.22) until the earliest completion of an executed operation or the earliest arrival of a moving AGV in the destination machine, whichever is earlier.

STEP 5. The terminal conditions

If all unloading operations are completed, then set $H = t$, calculate the schedule length $C_{max} = \sum_{t=1}^{H} T_t$ and terminate. Otherwise, set $c_{ijt} = c_{ijt-1} + T_t U_{ijt}$; $i \in I_j$; $j \in J$, calculate the in-process inventory f_{ijt}, g_{ijt} and location of idle vehicles a_{it} at the end of period t, set $t = t + 1$ and go back to *STEP 1*.

It should be pointed out that some priority rules (e.g., STP&TT) require ordering of all operation once at the beginning of the algorithm, whereas the other (e.g., SUP&TT) need H-times ordering at the beginning of each new assignment period t $(t = 1, \ldots, H)$.

The computational requirements are predominated by *STEP 2* to determine the variables V_{iljt} of the vehicle assignment for products transferring. *STEP 2* requires $O(m^2 n H)$ computations since at most n movements of products between m^2 pairs of machines are possible in every period. The computational complexity of the algorithm is dependent on the priority rules selected in *STEP 0*, and is varied from $O(m^2 n H + n \log n + m \log m)$ to $O(H(m^2 n + n \log n + m \log m))$, where the total number H of assignment periods is of the order of schedule length $O(C_{max})$.

7.2.5 Numerical examples

In this subsection the algorithm SMAV-ML is applied to determine detailed machine and vehicle scheduling for the two example problems described in Sect. 3.6.4.

Example 1. The FAS configuration for the first example is provided in Fig. 3.7. The system is composed of $m = 6$ assembly stations ($i = 1, 2, 3, 4, 5, 6$), each with a buffer of capacity $b_i = 8$ and a loading/unloading station L/U. The material handling system is bidirectional and consists of $a = 2$ AGVs with $q_o = 2$ time units required for a vehicle to move between any two neighbouring stations. The number of available pallets is $P = 10$.

The production batch consists of $v = 4$ selected product types with the requirements 2,4,5, and 6 units of product type 1,2,3, and 4, respectively. The graph of precedence relations is shown in Fig. 3.8 and the assembly times are given in Table 3.14. The product assignments for various loading objectives and different solution approaches are shown in Tables 3.14, 3.16, and 3.17 at the corresponding assembly times.

In the scheduling algorithm SMAV-ML the same dispatching rules have been used for the machine assignment and for the vehicle assignment (for both products transferring and idle movement). The computational experiments have been performed to compare all 30 combinations of 6 dispatching rules

178   7. Machine and Vehicle Scheduling in Flexible Assembly Systems

for machine and vehicle assignment and 5 dispatching rules for selection of destination machine.

For each set of product assignments Z_{ijk} presented in Tables 3.14, 3.16, and 3.17, Table 7.2 shows the schedule length $C_{max}^{(a)}$, $C_{max}^{(b)}$, $C_{max}^{(c)}$ obtained for each rule combination. The shortest schedules are achieved for operation assignments and product movements that minimize the maximum workload and total transportation time (case (b)). The average schedule length $C_{max}^{(c)}$ for loading and routing with $P_{max} = 72$ and $Q_{sum} = 208$ obtained using the sequential approach with the LP-based loading heuristic (case (c)) is not worse than that for the optimal loading with the same criteria (case (b)).

Table 7.2. Schedule length for different rule combinations

Rules for machine selection	Schedule length $C_{max}^{(a)}$, $C_{max}^{(b)}$, $C_{max}^{(c)}$				
	Dispatching rules for machine and vehicle assignment				
	$STP\&TT$	$SUP\&TT$	$LWKR$	$LUWKR$	$MWKR$	$MUWKR$
STT	210,206,200	208,194,214	202,204,222	214,204,200	208,210,198	220,204,188
SQS	216,194,198	199,180,221	202,181,234	206,195,207	194,210,206	222,183,198
LQS	190,207,200	202,202,194	220,202,208	220,194,224	197,207,206	218,206,210
$LW\&TT$	198,214,199	214,202,211	214,200,216	218,210,208	202,212,182	232,220,220
$LUW\&TT$	203,204,192	201,171,220	212,184,202	204,186,196	204,202,204	203,195,208

$C_{max}^{(a)}$ - schedule length for product assignments with $P_{max} = 60$ and $Q_{max} = 60$
$C_{max}^{(b)}$ - schedule length for product assignments with $P_{max} = 76$ and $Q_{sum} = 144$
$C_{max}^{(c)}$ - schedule length for product assignments with $P_{max} = 72$ and $Q_{sum} = 208$

In order to evaluate the effectiveness of the scheduling algorithm proposed and to compare the relative performance of alternative dispatching rules against the schedule length C_{max}, various computational experiments have been performed.

The results obtained indicate that good rule combinations are $SUP\&TT$ and SQS or $LUW\&TT$. The solutions are more sensitive to the dispatching rules for machine selection than to the rules for machine and vehicle assignment. The SQS and $LUW\&TT$ rules perform better than the other machine selection rules.

The above results cannot be considered to be general guides for selection of dispatching rules. Each particular case needs to be treated independently and the best combination of dispatching rules should be selected based on extensive simulation studies.

Example 2. The FAS configuration for the second example is provided in Fig. 3.9. The system is made up of $m = 3$ assembly stations ($i = 1, 2, 3$) and one loading/unloading station L/U. The available working space is identical for all stations: $b_1 = 10$, $b_2 = 10$, $b_3 = 10$.

The material handling system is bidirectional and consists of $a = 2$ AGVs with $q_o = 2$ time units required for an AGV to move between any two neighbouring stations.

The production batch consists of $v = 5$ products to be assembled of $n = 10$ types of components. The corresponding ordered sequences of tasks $j \in J_k$

required to assemble each product $k = 1, 2, 3, 4, 5$ are shown in Fig. 3.10 where L/U denotes loading/unloading operations. The loading and unloading times are identical for all products and are equal to 2 time units, i.e., $p_{L/U} = 2$.

The algorithm SMAV-ML is applied to determine machine and vehicle schedule for the example, given the task assignments and assembly routes determined by using the sequential approach with the LP-based loading heuristic. The selected assembly routes are shown in Fig. 3.11 and the solution values of the objective functions are $P_{max} = 44$ and $Q_{sum} = 60$, where the total transportation time includes transfer of each product from and to the L/U station.

Table 7.3. Schedule length for different rule combinations

Rules for machine selection	Schedule length C_{max}					
	Dispatching rules for machine and vehicle assignment					
	$STP\&TT$	$SUP\&TT$	$LWKR$	$LUWKR$	$MWKR$	$MUWKR$
STT	88	87	87	90	75	87
SQS	88	87	87	90	78	87
LQS	88	87	87	90	78	87
$LW\&TT$	88	87	87	90	78	87
$LUW\&TT$	88	87	87	90	75	87

In the scheduling algorithm SMAV-ML the same dispatching rules have been used for machine assignment and for vehicle assignment (for both products transferring and deadheading trips). The simulation studies have been performed to compare all 30 combinations of 6 dispatching rules for machine and vehicle assignment and 5 dispatching rules for selection of destination machine. Table 7.3 shows the schedule length C_{max} obtained for each rule combination. Note that the schedule length for the example depends on dispatching rules for machine and vehicle assignment and is nearly independent on the rule for selection of destination machine. Such a performance is due to the lack of duplicate task assignments in the solution of the loading problem, and hence only few assembly routes available for the small set of products in the example problem. When there are more assembly routes selected for a larger set of products, the schedule length is more clearly dependent on the rules for selection of destination machine, see Table 7.2 for the first example.

The shortest schedule with $C_{max} = 75$ is achieved for MWKR and STT or LUW&TT rule combinations. The machine and vehicle schedules obtained for this rule combination are shown in Fig. 7.6 and Fig. 7.7, respectively.

The machine schedule is made up of $H = 50$ assignment periods, each of duration T_t equal to 1 or 2 time units. In Fig. 7.6 for each assignment period t, its duration T_t and the pairs of operation and product numbers assigned to machines in this period are indicated. The product numbers $k = 1, 2, 3, 4, 5$ are replaced in Fig. 7.6 and Fig. 7.7 with letters A,B,C,D,E, respectively

The vehicle schedule in Fig. 7.7 shows a sequence of departure times of the two AGVs. For each departure time of a loaded or an empty vehicle,

					L	1	2	3	U
t =	1	T(t) =	2	L,E	--	--	--	--
t =	2	T(t) =	2	L,D	--	--	--	--
t =	3	T(t) =	2	L,C	1,E	--	--	--
t =	4	T(t) =	2	L,B	1,E	--	--	--
t =	5	T(t) =	2	L,A	3,E	--	--	--
t =	6	T(t) =	2	--	1,D	--	--	--
t =	7	T(t) =	2	--	1,D	5,E	--	--
t =	8	T(t) =	2	--	3,D	5,E	--	--
t =	9	T(t) =	2	--	2,C	--	--	--
t =	10	T(t) =	2	--	3,C	5,D	--	--
t =	11	T(t) =	2	--	6,E	5,D	--	--
t =	12	T(t) =	2	--	4,C	--	--	--
t =	13	T(t) =	2	--	6,D	7,E	--	--
t =	14	T(t) =	1	--	1,B	7,E	--	--
t =	15	T(t) =	1	--	1,B	5,C	--	--
t =	16	T(t) =	1	--	1,B	5,C	--	--
t =	17	T(t) =	1	--	1,B	5,C	8,E	--
t =	18	T(t) =	1	--	2,B	5,C	8,E	--
t =	19	T(t) =	1	--	2,B	--	8,E	--
t =	20	T(t) =	1	--	4,B	--	8,E	--
t =	21	T(t) =	1	--	4,B	--	8,E	--
t =	22	T(t) =	2	--	1,A	--	9,E	--
t =	23	T(t) =	2	--	1,A	--	7,D	--
t =	24	T(t) =	1	--	2,A	10,E	7,D	--
t =	25	T(t) =	1	--	2,A	10,E	8,D	--
t =	26	T(t) =	2	--	3,A	10,E	8,D	--
t =	27	T(t) =	2	--	4,A	5,B	8,D	--
t =	28	T(t) =	2	--	6,A	5,B	9,D	U,E
t =	29	T(t) =	2	--	--	--	7,C	--
t =	30	T(t) =	1	--	6,B	--	7,C	--
t =	31	T(t) =	1	--	6,B	--	8,C	--
t =	32	T(t) =	2	--	--	10,D	8,C	--
t =	33	T(t) =	2	--	--	10,D	8,C	--
t =	34	T(t) =	2	--	--	--	9,C	--
t =	35	T(t) =	2	--	--	--	7,B	--
t =	36	T(t) =	1	--	--	10,C	7,B	U,D
t =	37	T(t) =	1	--	--	10,C	9,B	U,D
t =	38	T(t) =	1	--	--	10,C	9,B	--
t =	39	T(t) =	1	--	--	10,C	8,A	--
t =	40	T(t) =	1	--	--	--	8,A	--
t =	41	T(t) =	1	--	--	--	8,A	--
t =	42	T(t) =	1	--	--	--	8,A	--
t =	43	T(t) =	1	--	--	10,B	8,A	--
t =	44	T(t) =	2	--	--	10,B	--	U,C
t =	45	T(t) =	1	--	--	10,B	--	--
t =	46	T(t) =	1	--	--	--	--	--
t =	47	T(t) =	1	--	--	--	--	U,A
t =	48	T(t) =	1	--	--	--	--	U,A
t =	49	T(t) =	1	--	--	--	--	--
t =	50	T(t) =	2	--	--	--	--	U,B

Fig. 7.6. Machine schedule for MWKR and STT rule combination

a pair of the sending and destination machines is indicated as well as the transferred product number, if a loaded trip is performed.

Finally, for a comparison the algorithm SMAV-ML is applied to determine machine and vehicle schedule for the example problem presented in Sect. 4.1.3 in which assembly sequences have been selected from among alternative sequences available for each product. The product assignments and assembly routes determined by using the sequential approach with the LP-based loading heuristic are shown in Fig. 4.4. The solution values of the objective functions are $P_{max} = 40$ and $Q_{sum} = 70$, where the total transportation time includes transfer of each product from and to the L/U station.

Table 7.4 shows the schedule length C_{max} obtained for various dispatching rule combinations considered in Table 7.3.

Table 7.4. Schedule length for different rule combinations – alternative assembly sequences

Rules for machine selection	Schedule length C_{max}					
	Dispatching rules for machine and vehicle assignment					
	$STP\&TT$	$SUP\&TT$	$LWKR$	$LUWKR$	$MWKR$	$MUWKR$
STT	87	74	78	82	80	80
SQS	87	75	85	82	85	80
LQS	87	75	81	82	84	80
$LW\&TT$	87	77	81	81	85	80
$LUW\&TT$	87	74	85	81	80	80

The shortest schedule with $C_{max} = 74$ is obtained for SUP&TT and STT or LUW&TT rule combinations. Unlike, for the example with single assembly sequences (Table 7.3), now the schedule length is also dependent on the rule for selection of destination machines. The duplicate task assignments and alternative assembly sequences exploited for selection of assembly routes at the loading level enable more destination machines to be selected during the scheduling procedure, and hence the deadheading trips to be chosen from a larger set of alternatives. As a result the schedule length depends on both dispatching rules for machine and vehicle assignment and dispatching rules for machine selection.

7.3 Machine and vehicle scheduling – a single-level approach

In this section a single-level algorithm for scheduling the release of products into the system, finished products unloading from the system and scheduling of machines and vehicles is presented. The algorithm, called SMAV-SL (Scheduling of Machines And Vehicles – Single-Level) [99, 100] is a list scheduling algorithm. Operations waiting for assignment to machines or vehicles are arranged in a list of operations ordered according to selected priority indices.

```
Time =  2: departure of AGV1 with product E from i = L/U  to l= 1
Time =  4: departure of AGV2 with product D from i = L/U  to l= 1
Time =  6: departure of empty    AGV1         from i = 1    to l= L/U
Time =  8: departure of AGV1 with product C from i = L/U  to l= 1
Time =  8: departure of empty    AGV2         from i = 1    to l= L/U
Time = 10: departure of AGV2 with product B from i = L/U  to l= 1
Time = 10: departure of AGV1 with product E from i = 1    to l= 2
Time = 12: departure of empty    AGV1         from i = 2    to l= L/U
Time = 14: departure of AGV1 with product A from i = L/U  to l= 1
Time = 16: departure of AGV1 with product D from i = 1    to l= 2
Time = 18: departure of empty    AGV2         from i = 1    to l= 2
Time = 18: departure of AGV1 with product E from i = 2    to l= 1
Time = 22: departure of AGV1 with product E from i = 1    to l= 2
Time = 22: departure of AGV2 with product D from i = 2    to l= 1
Time = 24: departure of AGV2 with product C from i = 1    to l= 2
Time = 26: departure of empty    AGV1         from i = 2    to l= 1
Time = 27: departure of AGV2 with product E from i = 2    to l= 3
Time = 28: departure of AGV1 with product D from i = 1    to l= 3
Time = 31: departure of empty    AGV2         from i = 3    to l= 2
Time = 33: departure of AGV2 with product C from i = 2    to l= 3
Time = 34: departure of empty    AGV1         from i = 3    to l= 1
Time = 36: departure of AGV2 with product E from i = 3    to l= 2
Time = 38: departure of AGV1 with product B from i = 1    to l= 2
Time = 42: departure of AGV2 with product E from i = 2    to l= L/U
Time = 46: departure of empty    AGV2         from i = L/U  to l= 3
Time = 46: departure of AGV1 with product B from i = 2    to l= 1
Time = 48: departure of AGV1 with product A from i = 1    to l= 3
Time = 48: departure of AGV2 with product D from i = 3    to l= 2
Time = 50: departure of empty    AGV2         from i = 2    to l= 1
Time = 52: departure of AGV2 with product B from i = 1    to l= 3
Time = 54: departure of empty    AGV1         from i = 3    to l= 2
Time = 56: departure of AGV1 with product D from i = 2    to l= L/U
Time = 56: departure of AGV2 with product C from i = 3    to l= 2
Time = 61: departure of empty    AGV2         from i = 2    to l= 3
Time = 62: departure of empty    AGV1         from i = L/U  to l= 2
Time = 63: departure of AGV2 with product B from i = 3    to l= 2
Time = 64: departure of AGV1 with product C from i = 2    to l= L/U
Time = 66: departure of empty    AGV2         from i = 2    to l= 3
Time = 68: departure of AGV2 with product A from i = 3    to l= L/U
Time = 69: departure of empty    AGV1         from i = L/U  to l= 2
Time = 71: departure of AGV1 with product B from i = 2    to l= L/U
```

Fig. 7.7. Vehicle schedule for MWKR and STT rule combination

7.3 Machine and vehicle scheduling – a single-level approach

7.3.1 Dispatching rules for machine and vehicle scheduling

In every period the assignment decisions are made using several dispatching rules. A priority rule is used to select next operation for assignment to an idle machine from the set of products waiting in its buffer. Similarly, priority rules are used to select products for transfer to other machines as well as to select the next movement of each loaded or empty vehicle.

Before an assignment decision is made, the products waiting for their next operations need ordering according to the priority index dependent on processing and transportation times. The processing time of an operation may depend on the machine selected for this operation. Similarly, transportation time depends on the pair of machines: sending and destination. Unlike the multi-level approach in which the product assignments and assembly routes are fixed at the loading level, and hence the processing times and transportation times for loaded trips are known, the single-level approach requires *mean values* of the corresponding time parameters to be used instead. The mean values are defined below.

$\bar{p}_j$ – mean processing time of operation j

$$\bar{p}_j = \sum_{i \in I_j} p_{ij}/|I_j|, \quad j \in J \tag{7.31}$$

$\bar{q}_j$ – mean transportation time for each product with completed operation j required for transferring it between machines to perform next operation $j+1$

$$\bar{q}_j = \sum_{i \in I_j} \sum_{l \in I_{j+1}} q_{il}/|I_j||I_{j+1}| \tag{7.32}$$

In order to describe different priority rules used in the single-level scheduling algorithm the following *auxiliary parameters* are introduced below:

$\overline{P}_j$ – total mean processing time of all operations j

$$\overline{P}_j = \bar{p}_j r_j \tag{7.33}$$

$\overline{P}_{jt}$ – mean processing time of operations j remaining at the beginning of period t

$$\overline{P}_{jt} = \bar{p}_j \left(r_j - \sum_{i \in I_j} \lceil c_{ijt-1}/p_{ij} \rceil \right) \tag{7.34}$$

where c_{ijt} is defined in (7.1)

$\overline{Q}_j$ – total mean transportation time required to transfer all products with completed operation j to perform next operation $j+1$

$$\overline{Q}_j = \bar{q}_j r_j \tag{7.35}$$

$\overline{Q}_{jt}$ – total mean transportation time $\overline{Q}_j$ calculated at the beginning of period t

$$\overline{Q}_{jt} = \overline{q}_j \left(r_j - \sum_{i \in I_j} \lfloor c_{ijt-1}/p_{ij} \rfloor \right) \quad (7.36)$$

$\overline{R}_j$ – total mean work remaining (processing and transportation time) required to complete all remaining operations beginning with j and ending with unloading operation of all products of a particular type

$$\overline{R}_j = \overline{P}_j + \overline{Q}_j + \overline{R}_{j+1}; \quad j \in J \setminus J_U \quad (7.37)$$

$$\overline{R}_j = \overline{P}_j; \quad j \in J_U \quad (7.38)$$

$\overline{R}_{jt}$ – total mean work remaining $\overline{R}_j$ calculated at the beginning of period t

$$\overline{R}_{jt} = \overline{P}_{jt} + \overline{Q}_{jt} + \overline{R}_{j+1t}; \quad j \in J \setminus J_U \quad (7.39)$$

$$\overline{R}_{jt} = \overline{P}_{jt}; \quad j \in J_U \quad (7.40)$$

The following priority rules are used to determine the assignment U_{ijt} of operations to machines, the assignment V_{iljt} of products (operations) to idle vehicles for transfer between machines as well as the selection of their next destination machines, and the assignment V_{ilt}^e of remaining idle vehicles to their next sending machines.

Dispatching rules for machine assignment.

STP&TT (Shortest Total Processing & Transportation Time):
$\min_j \{\overline{P}_j + \overline{Q}_j\}$
SUP&TT (Shortest Uncompleted Processing & Transportation Time):
$\min_j \{\overline{P}_{jt} + \overline{Q}_{jt}\}$
LWKR (Least Work Remaining): $\min_j \{\overline{R}_j\}$
LUWKR (Least Uncompleted Work Remaining): $\min_j \{\overline{R}_{jt}\}$
MWKR (Most Work Remaining): $\max_j \{\overline{R}_j\}$
MUWKR (Most Uncompleted Work Remaining): $\max_j \{\overline{R}_{jt}\}$

Dispatching rules for vehicle assignment for product transferring.

STP&TT (Shortest Total Processing & Transportation Time):
$\min_j \{\overline{Q}_j + \overline{P}_{j+1}\}$
SUP&TT (Shortest Uncompleted Processing & Transportation Time):
$\min_j \{\overline{Q}_{jt} + \overline{P}_{j+1t}\}$
LWKR (Least Work Remaining): $\min_j \{\overline{Q}_j + \overline{R}_{j+1}\}$
LUWKR (Least Uncompleted Work Remaining): $\min_j \{\overline{Q}_{jt} + \overline{R}_{j+1t}\}$
MWKR (Most Work Remaining): $\max_j \{\overline{Q}_j + \overline{R}_{j+1}\}$
MUWKR (Most Uncompleted Work Remaining): $\max_j \{\overline{Q}_{jt} + \overline{R}_{j+1t}\}$

Dispatching rules for machine selection.

STT (Shortest Transportation Time): $\min_l \{q_{il}\}$
SQS (Smallest Queue Size): $\min_l \{\sum_{j \in J} g_{ljt-1}\}$

Dispatching rules for idle vehicle assignment.

The assignment W_{ilt} of remaining idle vehicles to their next destination machines is made using the dispatching rules proposed above for vehicle assignment for product transferring. An idle vehicle waiting at machine i is send to such a machine l at which a product is waiting for transfer to perform its next operation $j+1$ with the highest priority.

7.3.2 Scheduling algorithm

The following three types of operations are distinguished in the single-level algorithm SMAV-SL for machine and vehicle scheduling:

– *Assembly* – denotes either an assembly operation at assembly station or a loading/unloading operation at L/U station,
– *Load_Trip*: – denotes a loaded trip of an AGV transferring product between machines,
– *Dead_Trip* – denotes a deadheading trip of an empty AGV between machines.

The following three types of *operation completion events* are defined for operations:

– *End of Assembly* – denotes the end of an assembly or a loading/unloading operation,
– *End of Load_Trip* – denotes the end of a loaded trip,
– *End of Dead_Trip* – denotes the end of a deadheading trip.

Each of the event of completion an operation denotes a change of state of some operation, and hence a change of the system state. Each operation can be in one of the following four states:

– *Passive* – denotes an assembly or unloading operation whose all predecessors have not yet been completed,
– *Active* – denotes an assembly or unloading operation whose all predecessors have been completed,
– *Performing* – denotes an operation that is being performed
– *Completed* – denotes an operation just completed.

Note that for *Assembly-* and *Load_Trip*-operations the state *Completed* does not imply an immediate beginning of the next operation. First, the corresponding product must be moved from machine or AGV to buffer.

An AGV staying at a station can be positioned at one of the two possible locations with respect to the station

– *At Station P/D Point* – at the station pickup/delivery point, where an AGV can be loaded or unloaded,
– *At Station* – at the station, but not at its P/D point.

It is assumed in the algorithm SMAV-SL that at most one AGV can be positioned *At Station P/D Point* at a time, and hence at most one product can be loaded to or unloaded from the station buffer at a time.

In the algorithm, each time an operation is started, the event of its completion is recorded in an ordered *list of events* with the operation completion time indicated. In every iteration the algorithm checks for the earliest event in the list of events. If the event leads to the release of a machine or an AGV, then an appropriate operation with the highest priority in the list of operations can be started.

Note that the states *End of Assembly* or *End of Load_Trip* does not imply an immediate beginning of the next operation by the station or the next trip by the AGV, respectively. First, the corresponding product must be moved from machine or AGV to a buffer.

Algorithm SMAV-SL

Starting
 – State of all loading operations – *Active*.
 – State of all remaining operations – *Passive*.
 – Locate all AGVs *At Station* L.
 – *List of events* consists of only one event with completion time 0.

Completion of operations corresponding to the earliest events
 1. *End of Assembly*:
    – Terminate completion of an assembly operation (*Completed*).
    – If there exists next assembly operation for the product, then make it *Active*.
    – If there exists an empty buffer at the station, then move the product from machine to the buffer.
 2. *End of Dead_Trip*:
    – Terminate *Dead_Trip* operation (*Completed*).
    – Position the empty AGV *At Station*.
 3. *End of Load_Trip*:
    – Terminate *Load_Trip* operation (*Completed*).
    – If location *At Station P/D Point* and at least one buffer are free, then position the AGV *At Station P/D Point* and move the product from AGV to the buffer. Otherwise, the event of completion this *Load_Trip* operation put again on the list of events with the completion time increased by 1 time unit.
    – Position AGV *At Station*.

Machine assignment for assembly or unloading operations
 – For each station select in its buffer an *Active* operation with the highest priority, waiting for assignment to this station.

- Move the product from buffer onto machine and begin performing the selected operation (*Performing*).
- The completion time of the performed operation put on the list of events.

Products loading in the system
- Select for loading the product with the highest priority of loading operation.
- If the loading station is idle, then begin the loading operation (*Performing*).
- The completion time of the performed loading operation put on the list of events.

Positioning empty vehicles *At Station P/D Points*
If location *At Station P/D Point* is free and there is an idle AGV *At Station* waiting for transferring a product, then position the AGV *At Station P/D Point* so that the product can be loaded on the AGV.

Vehicle assignment for product transferring and selection of destination stations
- In buffer of station with an idle AGV *At Station P/D Point* select product with the highest priority *Active* operation.
- Move the selected product from the buffer to the AGV, create a new *Load_Trip* operation, select the destination station with the highest priority and begin performing the operation (*Performing*).
- The completion time of the selected *Load_Trip* operation put on the list of events.

Selection of the sending stations for idle vehicles
- For idle AGVs find the sending stations having in their buffers products waiting for transfer to perform *Active* operations with the highest priority.
- Create new *Dead_Trip* operations and begin their performing. (*Performing*).
- The completion times of the selected *Dead_Trip* operations put on the list of events.

The End.

The solution results obtained by using the list scheduling algorithm **SMAV-SL** in the system with various types of operations and many different priority rules depends on the order in which various assignment decisions are made. The hierarchy of decision making has an effect on circumstances under which the system locking can occur. The possibility of occuring the system locking is minimized in algorithm **SMAV-SL** by an appropriate decision making hierarchy and by the procedure used to define start and completion of each type of operation.

7.3.3 Numerical examples

In this subsection the algorithm SMAV-SL is applied to determine detailed machine and vehicle schedule for the example problem similar to example 1 described in Sect. 3.6.4. In Sect. 7.2.5 the multi-level scheduling algorithm SMAV-ML has been applied to determine machine and vehicle schedules for a similar example using various combinations of dispatching rules.

The FAS configuration is provided in Fig. 3.7. The system is composed of $m = 6$ assembly stations ($i = 1, 2, 3, 4, 5, 6$), one loading station L and one unloading station U. Each station is equipped with a buffer allowing 4 pallets to be stored temporarily. The material handling system is bidirectional and consists of $a = 2$ AGVs with $q_o = 2$ time units required for a vehicle to move between any two neighbouring stations. The number of available pallets is $P = 10$.

The production batch consists of $v = 4$ selected product types with the requirements 2,4,5, and 6 units of product type 1,2,3, and 4, respectively.

$1 \longrightarrow 2 \longrightarrow 3 \longrightarrow 4$
$5 \longrightarrow 6 \longrightarrow 7 \longrightarrow 8 \longrightarrow 9$
$10 \longrightarrow 11 \longrightarrow 12 \longrightarrow 13 \longrightarrow 14 \longrightarrow 15 \longrightarrow 16$
$17 \longrightarrow 18 \longrightarrow 19 \longrightarrow 20$

Fig. 7.8. Graph of precedence relations

Graph of precedence relations for the example is shown in Fig. 7.8 where all operations are numbered from 1 to $n = 20$, including loading operations 1,5,10,17 and unloading operations 4,9,16,20. The new operations $j = 1, \ldots, 20$ denote the following pairs of the original assembly tasks and product types (see Fig. 3.8):
1=(j_L,1), 2=(4,1), 3=(6,1), 4=(j_U,1), 5=(j_L,2), 6=(3,2), 7=(5,2), 8=(8,2), 9=(j_U,2), 10=(j_L,3), 11=(1,3), 12=(2,3), 13=(4,3), 14=(7,3), 15=(8,3), 16=(j_U,3), 17=(j_L,4), 18=(1,4), 19=(8,4), 20=(j_U,4).

Table 7.5 shows the assembly times p_{ij} for the above defined operations.

Table 7.5. Assembly times

Product type k	1		2			3				4	
Operation j	2	3	6	7	8	11	12	13	14	15	18	19
Station												
$i = 1$	2	2				2	4	4	8	10		20
$i = 2$	4	3				3	2	8	10	10		20
$i = 3$	4	5				5	3	12	8	10		20
$i = 4$			5	2	12			8	10		4	20
$i = 5$			5	4	6			12	8		2	20
$i = 6$			3	1	12			4	10		5	20

7.3 Machine and vehicle scheduling – a single-level approach

The assembly and transportation schedules for the example are determined using algorithm SMAV-SL with the following combination of dispatching rules:

LWKR – for machine assignment,
LWKR – for vehicle assignment,
LWKR – for selection of sending machine for idle vehicle,
SQS – for selection of destination machine for loaded vehicle.

The resulting schedule length is $C_{max} = 175$.

The assembly schedules obtained for the example are shown in Tables 7.6 – 7.9 whereas the corresponding vehicle schedules for AGV1 and AGV2 are given in Tables 7.10 and 7.11.

The assembly schedules account for both the assembly and loading/unloading operations as well as product waiting times in buffers and product transfer times between the machines.

Table 7.6. Assembly schedule for products type 1

| Operation | | Time | | Location |
j	State	Start	End	Machine/Buffer/AGV
Assembly of 1st unit of product type 1				
1	Exe	115	117	L
2	Wait	117	118	L/B
2	Trans	118	120	L – 1/AGV1
2	Exe	120	122	1
3	Exe	122	124	1
4	Wait	124	131	1/B
4	Trans	131	133	1 – U/AGV2
4	Exe	133	135	U
Assembly of 2nd unit of product type 1				
1	Exe	120	122	L
2	Wait	122	125	L/B
2	Trans	125	129	L – 2/AGV2
2	Exe	129	133	2
3	Exe	133	136	2
4	Wait	136	142	2/B
4	Trans	142	146	2 – U/AGV1
4	Exe	146	148	U
Exe – performing of assembly operation,				
Wait – waiting in buffer B of machine i,				
Trans – transportation: machine i – machine l/AGV.				

Table 7.7. Assembly schedule for products type 2

Operation		Time		Location
j	State	Start	End	Machine/Buffer/AGV
Assembly of 1st unit of product type 2				
5	Exe	60	62	L
6	Wait	62	64	L/B
6	Trans	64	68	L – 5/AGV1
6	Exe	68	73	5
7	Exe	73	77	5
8	Exe	77	83	5
9	Wait	83	157	5/B
9	Trans	157	161	5 – U/AGV2
9	Exe	161	163	U
Assembly of 2nd unit of product type 2				
5	Exe	78	80	L
6	Wait	80	84	L/B
6	Trans	84	90	L – 4/AGV1
6	Wait	90	100	4/B
6	Exe	100	105	4
7	Exe	105	107	4
8	Exe	107	119	4
9	Trans	119	125	4 – U/AGV2
9	Exe	125	127	U
Assembly of 3rd unit of product type 2				
5	Exe	95	97	L
6	Wait	97	98	L/B
6	Trans	98	102	L – 5/AGV1
6	Exe	102	107	5
7	Exe	107	111	5
8	Exe	111	117	5
9	Wait	117	162	5/B
9	Trans	162	166	5 – U/AGV1
9	Exe	166	168	U
Assembly of 4th unit of product type 2				
5	Exe	100	102	L
6	Wait	102	106	L/B
6	Trans	106	108	L – 6/AGV1
6	Exe	108	111	6
7	Exe	111	112	6
8	Exe	112	124	6
9	Wait	124	171	6/B
9	Trans	171	173	6 – U/AGV2
9	Exe	173	175	U

Exe – performing of assembly operation,
Wait – waiting in buffer B of machine i,
Trans – transportation: machine i – machine l/AGV.

7.3 Machine and vehicle scheduling – a single-level approach 191

Table 7.8. Assembly schedule for products type 3

Operation j	State	Time Start	Time End	Location Machine/Buffer/AGV
\multicolumn{5}{c}{Assembly of 1st unit of product type 3}
10	Exe	L	2	L
11	Trans	2	4	L – 1/AGV1
11	Exe	4	6	1
12	Exe	6	10	1
13	Exe	10	14	1
14	Exe	14	22	1
15	Wait	22	40	1/B
15	Exe	40	50	1
16	Trans	50	53	1 – U/AGV1
16	Exe	52	54	U
\multicolumn{5}{c}{Assembly of 2nd unit of product type 3}
10	Exe	2	4	L
11	Trans	4	8	L – 2/AGV2
11	Exe	8	11	2
12	Exe	11	13	2
13	Exe	13	21	2
14	Exe	21	31	2
15	Wait	31	54	2/B
15	Exe	54	64	2
16	Wait	64	94	2/B
16	Trans	94	98	2 – U/AGV1
16	Exe	98	100	U
\multicolumn{5}{c}{Assembly of 3rd unit of product type 3}
10	Exe	4	6	L
11	Wait	6	8	L/B
11	Trans	8	14	L – 3/AGV1
11	Exe	14	19	3
12	Exe	19	22	3
13	Exe	22	34	3
14	Exe	34	42	3
15	Exe	42	52	3
16	Wait	52	107	3/B
16	Trans	107	113	3 – U/AGV2
16	Exe	113	115	U
\multicolumn{5}{c}{Assembly of 4th unit of product type 3}
10	Exe	6	8	L
11	Wait	8	18	L/B
11	Trans	18	22	L – 2/AGV2
11	Wait	22	31	2/B
11	Exe	31	34	2
12	Exe	34	36	2
13	Exe	36	44	2
14	Exe	44	54	2
15	Wait	54	64	2/B
15	Exe	64	74	2
16	Wait	74	89	2/B
16	Trans	89	93	2 – U/AGV2
16	Exe	93	95	U

Table 7.8 cont.

| Operation | | Time | | Location |
j	State	Start	End	Machine/Buffer/AGV
Assembly of 5th unit of product type 3				
10	Exe	8	10	L
11	Wait	10	14	L/B
11	Trans	14	16	L – 1/AGV2
11	Wait	16	22	1/B
11	Exe	22	24	1
12	Exe	24	28	1
13	Exe	28	32	1
14	Exe	32	40	1
15	Wait	40	50	1/B
15	Exe	50	60	1
16	Wait	60	74	1/B
16	Trans	74	76	1 – U/AGV1
16	Exe	76	78	U

Exe – performing of assembly operation,
Wait – waiting in buffer B of machine i,
Trans – transportation: machine i – machine l/AGV.

Table 7.9. Assembly schedule for products type 4

| Operation | | Time | | Location |
j	State	Start	End	Machine/Buffer/AGV
Assembly of 1st unit of product type 4				
17	Exe	10	12	L
18	Wait	12	37	L/B
18	Trans	37	39	L – 6/AGV1
18	Wait	39	43	6/B
18	Exe	43	48	6
19	Wait	48	68	6/B
19	Exe	68	88	6
20	Wait	88	167	6/B
20	Trans	167	169	6 – U/AGV2
20	Exe	169	171	U
Assembly of 2nd unit of product type 4				
17	Exe	12	13	L
18	Wait	14	39	L/B
18	Trans	40	45	L – 4/AGV2
18	Wait	46	51	4/B
18	Exe	52	55	4
19	Exe	56	75	4
20	Wait	76	111	4/B
20	Trans	112	117	4 – U/AGV1
20	Exe	118	119	U

Table 7.9 cont.

Operation j	State	Time Start	Time End	Location Machine/Buffer/AGV
\multicolumn{5}{c}{Assembly of 3rd unit of product type 4}			
17	Exe	14	16	L
18	Wait	16	22	L/B
18	Trans	22	28	L – 4/AGV1
18	Exe	28	32	4
19	Exe	32	52	4
20	Trans	52	58	4 – U/AGV2
20	Exe	58	60	U
\multicolumn{5}{c}{Assembly of 4th unit of product type 4}			
17	Exe	16	18	L
18	Wait	19	36	L/B
18	Trans	36	38	L – 6/AGV2
18	Exe	38	43	6
19	Wait	43	48	6/B
19	Exe	48	68	6
20	Wait	68	168	6/B
20	Trans	168	170	6 – U/AGV1
20	Wait	170	171	U/B
20	Exe	171	173	U
\multicolumn{5}{c}{Assembly of 5th unit of product type 4}			
17	Exe	19	21	L
18	Wait	24	28	L/B
18	Trans	28	32	L – 5/AGV2
18	Exe	32	34	5
19	Exe	34	54	5
20	Wait	54	154	5/B
20	Trans	154	158	5 – U/AGV1
20	Exe	158	160	U
\multicolumn{5}{c}{Assembly of 6th unit of product type 4}			
17	Exe	54	56	L
18	Wait	56	58	L/B
18	Trans	58	64	L – 4/AGV2
18	Wait	64	76	4/B
18	Exe	76	80	4
19	Exe	80	100	4
20	Wait	100	147	4/B
20	Trans	147	153	4 – U/AGV2
20	Exe	153	155	U

Exe – performing of assembly operation,
Wait – waiting in buffer B of machine i,
Trans – transportation: machine i – machine l/AGV.

The vehicle schedules account for two types of trips performed by vehicles: loaded trips and deadheading trips. In a vehicle schedule each loaded trip is described by the number of active operation which requires the corresponding product to be transferred between machines, a pair of the machines: i – sending and l – destination, as well as the vehicle departure time from the sending machine and its arrival time in the destination machine, measured by time elapsed from the beginning of the assembly process. Similarly, each deadheading trip is described by the pair of machines and the departure and arrival times.

Table 7.10. Time table for AGV 1

Trip		Time		Machine	
$j/i-l$	Type	Start	End	Sending i	Destination l
11	Loaded	2	4	L	1
1–L	Deadheading	6	8	1	L
11	Loaded	8	14	L	3
3–L	Deadheading	16	22	3	L
18	Loaded	22	28	L	4
4–L	Deadheading	31	37	4	L
18	Loaded	37	39	L	6
6–1	Deadheading	39	43	6	1
16	Loaded	50	52	1	U
6	Loaded	64	68	L	5
5–1	Deadheading	68	74	5	1
16	Loaded	74	76	1	U
6	Loaded	84	90	L	4
4–2	Deadheading	90	94	4	2
16	Loaded	94	98	2	U
6	Loaded	98	102	L	5
5–L	Deadheading	102	106	5	L
6	Loaded	106	108	L	6
6–4	Deadheading	108	112	6	4
20	Loaded	112	118	4	U
2	Loaded	118	120	L	1
1–2	Deadheading	138	142	1	2
4	Loaded	142	146	2	U
U–5	Deadheading	150	154	U	5
20	Loaded	154	158	5	U
U–5	Deadheading	158	162	U	5
14	Loaded	162	166	5	U
U–6	Deadheading	166	168	U	6
20	Loaded	168	170	6	U
Loaded – a loaded trip of an AGV between machines, Deadheading – an empty trip of an AGV between machines.					

7.3 Machine and vehicle scheduling – a single-level approach

Table 7.11. Time table for AGV 2

Trip		Time		Machine	
$j/i-l$	Type	Start	End	Sending i	Destination l
11	Loaded	4	8	L	2
2–L	Deadheading	10	14	2	L
11	Loaded	14	15	L	1
1–L	Deadheading	16	18	1	L
11	Loaded	18	22	L	2
2–L	Deadheading	24	28	2	L
18	Loaded	28	32	L	5
5–L	Deadheading	32	36	5	L
18	Loaded	36	38	L	6
6–L	Deadheading	38	40	6	L
18	Loaded	40	46	L	4
19	Loaded	52	58	4	U
18	Loaded	58	64	L	4
4–2	Deadheading	85	89	4	2
16	Loaded	89	93	2	U
U–3	Deadheading	101	107	U	3
16	Loaded	107	113	3	U
U–4	Deadheading	113	119	U	4
9	Loaded	119	125	4	U
2	Loaded	125	129	L	2
2–1	Deadheading	129	131	2	1
4	Loaded	131	133	1	U
U–4	Deadheading	141	147	U	4
20	Loaded	147	153	4	U
U–5	Deadheading	153	157	U	5
9	Loaded	157	161	5	U
U–6	Deadheading	165	167	U	6
20	Loaded	167	169	6	U
U–6	Deadheading	169	171	U	6
9	Loaded	171	173	6	U

Loaded – a loaded trip of an AGV between machines,
Deadheading – an empty trip of an AGV between machines.

The assembly schedules obtained for the example indicate that each product is completely assembled at one station, and hence no loading trips between stations (excluding trips from/to L/U stations) are performed (see Tables 7.10, 7.11).

References

1. Agnetis, A., Arbib, C., Lucertini, M., Nicolo, F. (1990): Part routing in flexible assembly systems. *IEEE Transaction on Robotics and Automation.* **6**, 697–705
2. Agnetis, A. (1996): Planning the routing mix in FASs to minimize total transportation time. *International Journal of Flexible Manufacturing Systems.* **7**(3)
3. Ahmadi, J., Grotzinger, S., Johnson, D. (1988): Component allocation and positioning for a dual delivery placement machine. *Operations Research.* **36**, 176-191
4. Ahmadi, J., Grotzinger, S., Johnson, D. (1991): Emulating concurrency in a circuit card assembly system. *International Journal of Flexible Manufacturing Systems.* **3**, 45-70
5. Ahmadi, J., Ahmadi, R., Matsuo, H., Tirupati, D. (1995): Component fixture positioning/sequencing for printed circuit board assembly with concurrent operations. *Operations Research.* **43**, 444-457
6. Ahuja, R.K., Magnanti, T.L., Orlin, J.B. (1993): *Network Flows.* Prentice-Hall, Englewood Cliffs, New Jersey
7. Ammons, J.C., Lofgren, C.B., McGinnis, L.F. (1985): A large scale machine loading problem in flexible assembly. *Annals of Operations Research.* **3**, 319–332
8. Andreasen, M.M., Ahm, T. (1988): *Flexible Assembly Systems.* IFS Publications, Springer-Verlag, Berlin
9. Arbib, C., Lucertini, M., Nicolo, F. (1990): Workload balance and part-transfer minimization in flexible manufacturing systems. *International Journal of Flexible Manufacturing Systems.* **3**, 5–25
10. Ball, M.O., Magazine, M.J. (1988): Sequencing of insertion in printed circuit boards assembly. *Operations Research.* **36**, 192-201
11. Bard, J.F. (1985): Geometric and algorithmic developments for a hierarchical planning problem. *European Journal of Operational Research.* **19**, 372–383.
12. Bard, J.F. (1989): Assembly line balancing with parallel workstation and dead time. *International Journal of Production Research.* **27**, 1005-1018
13. Bard, J.F., Clayton, R.W., Feo, T.A. (1994): Machine setup and component placement in printed circuit board assembly. *International Journal of Flexible Manufacturing Systems.* **6**, 5-31
14. Bard, J.F., Dar-El, E., Shtub, A. (1992): An analytic framework for sequencing mixed model assembly lines. *International Journal of Production Research.* **30**, 35-48
15. Bedworth, D.D., Henderson, M.R., Wolfe, P.M. (1991): *Computer-Integrated Design and Manufacturing.* McGraw-Hill, New York
16. Berrada, M., Stecke, K.E. (1986): A branch and bound approach for machine load balancing in flexible manufacturing systems. *Management Science.* **32**, 1316–1335

17. Bilge, U., Ulusoy, G. (1995): A time window approach to simultaneous scheduling of machines and material handling system in an FMS. *Operations Research.*, **43**, 1058–1070
18. Błażewicz, J., Eiselt, A., Finke, G., Laporte, G., Weglarz, J. (1991): Scheduling tasks and vehicles in a flexible manufacturing systems. *International Journal of Flexible Manufacturing Systems.* **4**, 5–16
19. Błażewicz, J., Ecker, K.H., Schmidt, G., Weglarz, J. (1994): *Scheduling in Computer and Manufacturing Systems.* Springer-Verlag, Berlin
20. Błażewicz, J, Burkard, R., Finke, G. Woeginger, G. (1994): Vehicle scheduling in two-cycle flexible manufacturing systems. *Mathematical and Computer Modelling.* **20**, 19–31
21. Bonneville, F., Perrard, C., Henrioud, J.M. (1995): A genetic algorithm to generate and evaluate assembly plans. In: *Proceedings of the INRIA/IEEE Symposium on Emerging Technologies and Factory Automation.* Paris, October 10-13, **3**, 231–239
22. Boothroyd, G., Poli, C., Murch, L.E. (1982): *Automatic Assembly.* Marcel Dekker, New York
23. Boothroyd, G., Dewhurst, P. (1988): Product design for manufacture and assembly. *Manufacturing Engineering.* No. 4, 42-46
24. Boothroyd, G., Dewhurst, P. (1991): *Product Design for Assembly.* Boothroyd Dewhurst Inc., USA
25. Brah, S.A., Hunsucker, J.L. (1991): Branch and bound algorithm for the flow shop with multiple processors. *European Journal of Operational Research.* **51**, 88–99
26. Crama, Y., Oerlemans, A.G., Spieksma, F.C.R. (1994): *Production Planning in Automated Manufacturing.* Lecture Notes in Economics and Mathematical Systems 414, Springer-Verlag, Berlin
27. Dannenbring, D.G. (1977): An evaluation of flow shop sequencing heuristics. *Management Science.* **23**, 1174–1182
28. Delchambre, A. (1992): *Computer-Aided Assembly Planning.* Chapman & Hall, London
29. Dewhurst, P., Boothroyd, G. (1983): Computer-aided design for assembly. *Assembly Engineering.* no. 2, 18-22
30. Drexl, A., Kimms, A. (1997): Sequencing JIT mixed-model assembly lines under station load- and part usage-constraints. Paper presented at the Workshop on Scheduling in Computer and Manufacturing Systems, Dagstuhl, Germany, June 1997
31. Dutta, S.K., Cunningham, A.A. (1975): Sequencing two-machine flow-shops with finite intermediate storage. *Management Science.* **21**, 989–996
32. Egbelu, P.J., Tanchoco, J.M. (1984): Characterization of automated guided vehicle dispatching rules. *International Journal of Production Research.* **22**, 359–374
33. Famos-Eureka Project EU 72 (1987): *Preliminary study report on European collaboration in the field of flexible automated assembly systems.* Department of Trade and Industry, London, June 1987
34. Famos: the Factory of the Future, (1991): *Eureka News.* No. 12, April 1991, 4-9
35. Feo, T.A., Bard, J.F., Holland, S.D. (1995): Facility-wide planning and scheduling of printed wiring board assembly. *Operations Research.* **43**, 219-230
36. Fujimoto, H., Yasuda, K., Iwahasi, K. (1994): Simulation analysis of decision rule sets for multilevel production scheduling in flexible manufacturing systems. In: *Proceedings of 1994 Japan-USA Symposium on Flexible Automation.* Kobe, July 11-18, 811–814

37. Gabbani, D., Magazine, M. (1986): An interactive heuristic approach for multiobjective integer-programming problems. *Journal of Operational Research Society.* **37**, 285–291
38. Gangadharan, R., Rajendran, C. (1993): Heuristic algorithms for scheduling in the no-wait flowshop. *International Journal of Production Economics.* **32**, 285–290
39. Ghosh, S., Gagnon, R.J. (1989): A comprehensive literature review and analysis of the design, balancing and scheduling of assembly systems. *International Journal of Production Research.* **27**, 637–670
40. Glover, F., Woolsey, E. (1974): Converting the 0-1 polynomial programming problem to a 0-1 linear program. *Operations Research.* **22**, 180–182
41. Graves, S.C., Lamar, B.W. (1983): An integer programming procedure for assembly system design problems. *Operations Research.* **31**, 522-545
42. Graves, S.C., Redfield, C.H. (1988): Equipment selection and task assignment for multiproduct assembly system design. *International Journal of Flexible Manufacturing Systems.* **1**(1), 31–50
43. Greene, J.T., Sadowski, P.R. (1986): A mixed integer program for loading and scheduling multiple flexible manufacturing cells. *European Journal of Operational Research.* **24**, 379–386
44. Gupta, J.N.D., Tunc, E.A. (1991): Schedules for a two-stage hybrid flowshop with parallel machines at the second stage. *International Journal of Production Research.* **29**, 1489–1502
45. Gupta, Y.P., Gupta, M.C., Bector, C.R. (1989): A review of scheduling rules in flexible manufacturing systems. *International Journal of Computer Integrated Manufacturing.* **2**, 356–377
46. Hall, D.N., Stecke, K.E. (1986): Design problems of flexible assembly systems. In: K.E.Stecke and R.Suri (eds.), *Proceedings of the Second ORSA/TIMS Conference on Flexible Manufacturing Systems: Operations Research Models and Applications.* Elsevier, Amsterdam, 145-156
47. Hall, R. (1983): *Zero Inventories.* Dow Jones-Irwin, Homewood, Illinois
48. He, W.D., Kusiak, A. (1994): Design of products and assembly systems for agility. *Proceedings of the International Mechanical Engineering Congress and Exposition.* ASME, New York, PED-Vol. 68-1, 29-34
49. Hunsucker, J.L., Shah, J.R. (1994): Comparative performance analysis of priority rules in a constrained flow shop with multiple processors environment. *European Journal of Operational Research.* **72**, 102–114
50. Jiang, J., Hsiao, W. (1994): Mathematical programming for the scheduling problem with alternate process plans in FMS. *Computers and Industrial Engineering.* **27**(10), 15–18
51. Johnson, S.M. (1954): Optimal two- and three-stage production schedules with setup times included. *Naval Research Logistics Quarterly.* **1**, 61–68
52. Kim, Y.-D., Yano, C.A. (1994): A new branch and bound algorithm for loading problems in flexible manufacturing systems. *International Journal of Flexible Manufacturing Systems.* **6**, 361–382
53. Kirkavak, N., Dincer, C. (1993): Analytical loading models in flexible manufacturing systems. *European Journal of Operational Research.* **71**, 17–31
54. Klundert, J.J. van de (1996): *Scheduling Problems in Automated Manufacturing.* Dissertation no. 96-35, University of Limburg, Mastricht
55. Korcyl, A., Lebkowski, P., Sawik, T. (1995): Selection of assembly sequences and balancing workloads in a flexible assembly line. In: *Proceedings of the INRIA/IEEE Symposium on Emerging Technologies and Factory Automation.* Paris, October 10-13, **3**, 349–359

56. Kubiak, W. (1993): Minimizing variation of production rates in just-in-time systems: A survey. *European Journal of Operational Research.* **66**, 259–271
57. Kubiak, W., Sethi, S. (1991): A note on "Level schedules for mixed-model assembly lines in just-in-time production systems. *Management Science.* **37**(1), 121–122
58. Kubiak, W., Sethi, S. (1994): Optimal just-in-time schedules for flexible transfer lines. *International Journal of Flexible Manufacturing Systems.* **6**, 137–154
59. Kubiak, W., Steiner, G., Yeomans, J.S. (1997): Optimal level schedules for mixed-model, multi-level just-in-time assembly systems. *Annals of Operations Research.* **69**, 241–259
60. Kusiak, A. (1990): *Intelligent Manufacturing Systems.* Prentice Hall, Englewood Cliffs, New Jersey
61. Kusiak, A. (1994): Concurrent engineering: issues, models, and solution approaches. In: R.Dorf and A.Kusiak (eds.), *Handbook of Design, Manufacturing and Automation.* John Wiley & Sons, New York, 35-49
62. Laarhoven, P.J.M.van, Zijm, W.H.M. (1993): Production preparation and numerical control in PCB assembly. *International Journal of Flexible Manufacturing Systems.* **5**, 187-207
63. Lee, H.F., Johnson, R.V. (1991): A line-balancing strategy for designing flexible assembly systems. *International Journal of Flexible Manufacturing Systems.* **3**, 91–120
64. Lee, H.F., Stecke, K.E. (1996): An integrated design support method for flexible assembly systems. *Journal of Manufacturing Systems.* **15**(1), 13–32
65. Leisten, R. (1990): Flowshop sequencing problems with limited buffer storage. *International Journal of Production Research.* **28**, 2085–2100
66. Lenstra, J., Shmoys, D.B., Tardos, E. (1990): Approximation algorithms for scheduling unrelated parallel machines. *Mathematical Programming.* **46**, 259–271
67. Liu, C-M, Duh, S-H. (1992): Study of AGVS design and dispatching rules by analytical and simulation methods. *International Journal of Computer Integrated Manufacturing.* **5**, 290–299
68. Maimon, O.Z., Nof, S.Y. (1985): Coordination of robots sharing assembly tasks. *ASME Journal of Dynamic Systems Measurements and Control.* **107**, 292-307
69. Martello, S., Toth, P. (1990): *Knapsack Problems: Algorithms and Computer Implementations.* John Wiley & Sons, Chichester, England
70. Miltenburg, J. (1989): Level schedules for mixed-model assembly lines in just-in-time production systems. *Management Science.* **32**(2), 192–207
71. Miltenburg, J., Sinnamon, G. (1989): Scheduling mixed-model multilevel assembly lines in just-in-time production systems *International Journal of Production Research.* **27**, 1487–1509
72. Miltenburg, J., Steiner, G., Yeomans, S. (1990): A dynamic programming algorithm for scheduling mixed-model, just-in-time production systems. *Mathematical and Computer Modelling.* **13**(3), 57–66
73. Modi, B.K., Shanker, K. (1994): A formulation and solution methodology for part movement minimization and workload balancing at loading decisions in FMS. *International Journal of Production Economics.* **34**(1), 73–82
74. Monden, Y. (1983): *Toyota Production System.* Institute of Industrial Engineers Press, Norcross, Georgia
75. Montazeri, M., Van Wassenhove, L.N. (1990): Analysis of scheduling rules for an FMS. *International Journal of Production Research.* **28**, 785–802
76. Narasimhan, S.L., Panwalkar, S.S. (1984): Scheduling in a two-stage manufacturing process. *International Journal of Production Research.* **22**, 555–564

77. Narasimhan, S.L., Mangiameli, P.M. (1987): A comparison of sequencing rules for a two-stage hybrid flow shop. *Decision Sciences*. **18**, 250–265
78. Nemhauser, G.L., Wolsey, L.A. (1988): *Integer and Combinatorial Optimization*. John Wiley & Sons, New York, 1988
79. Nevins, J.L ., Whitney, D.E. (1989): *Concurrent Design of Products & Processes*. McGraw-Hill, New York
80. Papadimitriou, C.H., Steiglitz, K. (1982): *Combinatorial Optimization: Algorithms and Complexity*. Prentice-Hall, Englewood Cliffs, NJ
81. Pinedo, M. (1992): Scheduling of flexible assembly system. In: Kusiak, A. (ed.), *Intelligent Design and Manufacturing*. John Wiley & Sons, New York, 449–468
82. Potts, C.N., Sevastjanov, S.V., Strusevich, V.A., Van Wassenhove, L.N., Zwaneveld, C.M. (1995): The two-stage assembly scheduling problem – complexity and approximation. *Operations Research*. **43**, 346–355
83. Rabinowitz, G., Mehrez, A., Samaddar, S. (1991): A scheduling model for multi-robot assembly cells. *International Journal of Flexible Manufacturing Systems*. **3**, 149-180
84. Rachamadugu, R., Stecke, K.E. (1994): Classification and review of FMS scheduling procedures. *Production Planning and Control*. **5**, 2–20
85. Raman, N. Talbot, B., Rachamadugu, R.V. (1986): Simultaneous scheduling of machines and material handling devices in automated manufacturing. In: Stecke, K.E., Suri, R.(eds.), *Proceedings of the Second ORSA/TIMS Conference on Flexible Manufacturing Systems: Operations Research Models and Applications*. Elsevier, Amsterdam, 321–332
86. Ravi, T., Lashkari, R.S., Dutta, S.P. (1991): Selection of scheduling rules in FMSs – A simulation approach. *International Journal of Advanced Manufacturing Technology*. **6**, 246–262
87. Sabuncuoglu, I., Hommertzheim, D.L. (1992): Dynamic dispatching algorithm for scheduling machines and automated guided vehicles in a flexible manufacturing system. *International Journal of Production Research*. **30**, 1059–1079
88. Sabuncuoglu, I., Hommertzheim, D.L. (1992): Experimental investigation of FMS machine and AGV scheduling rules against the mean flow- time criterion. *International Journal of Production Research*. **30**, 1617–1635
89. Santos, D.L., Hunsucker, J.L., Deal, D.E. (1995): Global lower bounds for flow shops with multiple processors. *European Journal of Operational Research*. **80**, 112–120
90. Sawik, T. (1982): Hierarchical scheduling two-stage multi-machine production with finite intermediate storage. *UMM Scientific Bulletins, Automatics*. **32**, 373–383
91. Sawik, T. (1987): Multilevel scheduling of multistage production with limited in-process inventory. *Journal of the Operational Research Society*. **38**, 651–664
92. Sawik, T. (1988): Hierarchical scheduling flowshops with parallel machines and finite in-process buffers. *Archiwum Automatyki i Telemechaniki*. **33**, 403–414
93. Sawik, T. (1988): Scheduling flow-shops with parallel machines and finite in-process buffers by multilevel programming. In: Iri, M, Yajima, K. (eds.) *System Modelling and Optimization*. Springer-Verlag, Berlin, 691–700
94. Sawik, T. (1989): Operation scheduling and vehicle routing in an FMS. In: *Proceedings of the Conference on the Practice and Theory of Operations Management*. AFCET, Paris, 31–38
95. Sawik, T. (1990): Modelling and scheduling of a flexible manufacturing system. *European Journal of Operational Research*. **45**, 177–190
96. Sawik, T. (1991): Task allocation, operations scheduling and vehicle routing in FMSs. *Archiwum Automatyki i Robotyki*. **36**, 19-38

97. Sawik, T. (1992): *Optymalizacja dyskretna w elastycznych systemach producyjnych.* (Discrete Optimization in Flexible Manufacturing Systems). (in Polish with English summary), WNT Publishers, Warszawa
98. Sawik, T. (1993): A scheduling algorithm for flexible flow lines with limited intermediate buffers. *Applied Stochastic Models and Data Analysis.* Special issue on Manufacturing Systems. **9**, 127–138
99. Sawik, T. (1993): Scheduling of machines and vehicles in an FMS: single-level versus multi-level approach. *UMM Scientific Bulletins, Automatics.* **64**(4), 541–568
100. Sawik, T. (1994): Algorithms for simultaneous scheduling of machines and vehicles in a FMS. In: Cohen, G., Quadrat, J.-P. (eds.) *Discrete Event Systems*, Lecture Notes in Information Sciences 199, Springer-Verlag, 616–621
101. Sawik, T. (1994): New algorithms for scheduling flexible flow lines. In: *Proceedings of Japan-U.S.A. Symposium on Flexible Automation.* Kobe, July 11-18, **3**, 1091–1096
102. Sawik, T. (1995): Integer programming models for the design and balancing of flexible assembly systems. *Mathematical and Computer Modelling.* **21**(4), 1–12
103. Sawik, T. (1995): Balancing machine workload and part movement in a flexible manufacturing system. *UMM Scientific Bulletins, Electrotechnics.* No. 4, 357–367
104. Sawik, T. (1995): Scheduling flexible flow lines with no in-process buffers. *International Journal of Production Research.* **33**, 1359–1370
105. Sawik, T. (1995): Dispatching scheduling of machines and vehicles in a flexible manufacturing system. In: *Proceedings of the INRIA/IEEE Symposium on Emerging Technologies and Factory Automation.* Paris, October 10-13, **2**, 3–13
106. Sawik, T. (1996): Sequential loading and routing in a FAS. In: *Proceedings of Rensselaer's Fifth International Conference on Computer Integrated Manufacturing & Automation Technology.* Grenoble, May 29-31, 48–53
107. Sawik, T. (1996): A two-level heuristic for machine loading and assembly routing in a FAS. In: *Proceedings of the 5th IEEE International Conference on Emerging Technologies and Factory Automation.* Kauai, HI November 18-21, 143–149
108. Sawik, T. (1996): A multilevel machine and vehicle scheduling in a flexible manufacturing system. *Mathematical & Computer Modelling.* **23**(7), 45–57
109. Sawik, T. (1997): Flexible assembly line balancing with alternate assembly plans and duplicate task assignments. In: *Proceedings of the 6th IEEE International Conference on Emerging Technologies and Factory Automation.* UCLA, CA September 9-12, 171–176
110. Sawik, T. (1997): An interactive approach to bicriterion loading of a flexible assembly system. *Mathematical and Computer Modelling.* **24**(6), 71–83
111. Sawik, T. (1997): An LP-based approach for loading and routing in a flexible assembly system. In: *Proceedings of International Conference on Industrial Engineering and Production Management.* Lyon, October 20-24, 216–225
112. Sawik, T. (1998): A lexicographic approach to bi-objective loading of a flexible assembly system. *European Journal of Operational Research.* **107**(3)
113. Sawik, T. (1998): Simultaneous loading, routing and assembly plan selection in a flexible assembly system. *Mathematical and Computer Modelling.* (forthcoming)
114. Schneeweiss, Ch. (1995): Hierarchical structures in organizations: A conceptual framework. *European Journal of Operational Research.* **86**, 4-31
115. Schrage, L., Cunningham, K. (1991): *LINGO, Optimization Modeling Language.* LINDO Systems Inc., Chicago

116. Singh, N. (1995): *Systems Approach to Computer Integrated Design and Manufacturing*. John Wiley & Sons, Inc. New York
117. Sinriech, D. (1995): Network design models for discrete material flow systems: A literature review. *International Journal of Advanced Manufacturing Technology.* **10**, 277-291
118. Sriskandarajah, C., Sethi, S.P. (1989): Scheduling algorithms for flexible flowshops: Worst and average case performance. *European Journal of Operational Research.* **43**, 143-160
119. Stecke, K.E. (1985): Design, planning, scheduling, and control problems of flexible manufacturing systems. Annals of Operations Research. **3**, 3-12
120. Stecke, K.E., Raman, N. (1995): FMS planning decisions, operating flexibilities and system performance. IEEE Transactions on Engineering Management.
121. Stecke, K.E., Solberg, J.J. (1985): The optimality of unbalancing both workloads and machine group sizes in closed queueing networks of multi-server queues. *Operations Research.* **33**(4), 882-910
122. Steuer, R.E. (1986): *Multiple Criteria Optimization: Theory, Computation, and Applications.* John Wiley & Sons, New York
123. Talbot, F.B., Paterson, J.H. (1984): An integer programming algorithm with network cuts for solving the assembly line balancing problem. *Management Science.* **30**(1), 85-99
124. Tang, L., Yih, Y., Liu, C. (1993): A study on decision rules of a scheduling model in an FMS. *Computers in Industry.* **22**, 1-13
125. Ulusoy, G., Bilge, U. (1993): Simultaneous scheduling of machines and automated guided vehicles. *International Journal of Production Research.* **31**, 2857-2873
126. Whitney, D.E. (1985): Planning programmable assembly systems. In: S.Y. Nof (ed.), *Handbook of Industrial Robotics*, John Wiley & Sons, New York
127. Winters, I.J., Burstein, M.C. (1992): A concurrent development tool for flexible assembly systems. *International Journal of Flexible Manufacturing Systems.* **4**, 293-307
128. Wittrock, R.J. (1985): Scheduling algorithms for flexible flow lines. *IBM Journal of Research and Development.* **29**, 401-412
129. Wittrock, R.J. (1988): An adaptable scheduling algorithm for flexible flow lines. *Operations Research.* **36**, 445-453

Index

0 − 1 programming, *see* Mathematical programming

Agile assembly   20
Algorithms:
- **RITM**   123
- **RITM-NS**   134
- **SMAV-ML**   176
- **SMAV-SL**   186

Assembly liaisons   31
Assembly plan
  (*see also* Assembly sequence) 20, 30, 43–44
- selection   87
Assembly plans   17
- alternative   87, 89, 94
- treelike   87–88
Assembly robots   2, 19, 44, 107
Assembly route   35, 41, 132
Assembly routes   36, 62–63, 70, 72, 83, 164–165
- alternative   43, 49, 56, 58, 63, 91, 99
- fixed   43, 56–58, 62
Assembly routing   36, 164
Assembly schedule   122–125, 135–137, 189–192
Assembly sequence   31–32, 42
Assembly sequences   33–34, 94
- alternative   102, 181
- graph of   96, 103
Assembly station   5, 44
Assignment period   168–169, 174, 177, 179
Assignment problem   146–147
Automated Guided Vehicles (AGVs)
  5, 22, 37–38, 43, 165–167, 177–179, 188
Automated assembly   27
- design for   27

Balanced scheduling   148–149

Balancing station workloads   47, 49, 57–59, 70, 91, 99, 100
Bicriterion machine loading   87
- and product routing   57, 69
- , product routing and assembly plan selection   91, 99
Blocking time   23
Buffers   22, 26, 37–38, 118, 122–124, 139, 152–153, 163, 189

Car sequencing   143
Component retrieval problem   40, 114–115
Computational complexity   124, 128, 135, 158, 177
Cut constraint   104

Deadheading trips   22, 167, 181, 194
Dispatching algorithms   165
Dispatching rules   165, 178–179, 189
- for idle vehicle assignment   176, 185
- for machine assignment   175, 184
- for machine selection   175, 185
- for vehicle assignment   175, 184
Duplicate task assignments
  (*see also* Task assignments) 42, 49–50, 87, 98, 181
Dynamic programming   151

Flexibility capacity   19, 44–50
Flexible assembly   17
- cell   6, 8
- line   6–8, 47, 50, 97–101, 117–118, 130–131, 143, 148, 152–153, 157
- - design and balancing   50, 51
- - with parallel machines   45, 47, 51
- system (FAS)   1, 5, 17–19, 35–37, 43, 163
- - balancing   36, 42–44
- - design   17–19, 43
- - planning   17, 32, 35

– – scheduling   35, 38, 163
– – – and control   17

Heuristics:
– interactive   60
– linear relaxation
  (*see also* LP-based)   75, 79, 83, 92, 101
– part-by-part   117, 121-122
– period-by-period   156, 172
– single-pass   122, 130, 139
Hierarchical scheduling   156

Integer programming, *see* Mathematical programming

Just -In -Time (JIT)   118, 142–143
– assembly line   142–143
– scheduling   142–144

LINGO   54–56, 68, 96, 103
Loaded trips   167, 194
Lot size scheduling   153
Lower bound   62, 67, 124, 135, 139, 140–142, 158
LP-based heuristic   76, 79, 178, 181

Machine and vehicle scheduling   36, 164–165, 169, 177
Machine assignment   154, 160, 169–170, 172, 176, 186
Machine blocking   19, 37, 120–121, 127, 137, 139
Machine loading   35–36, 164
Machine schedule   164–165, 179–180
Machine workload   119, 124–125, 135
Makespan
  (*see also* Schedule length)   117, 120–121, 124, 126, 132, 139, 142, 156, 164
Material flow networks   23
Material handling system   17, 22, 35, 72, 94, 102, 177–178, 188
Mathematical programming:
– 0 − 1 integer   90–91, 97–99
– bi-objective integer   42, 56, 91
– integer   42–43
– mixed-integer   35, 113
– multilevel   153, 155, 161
– nonlinear integer   149
– quadratic integer   115
Mean values   183
Mechanical assembly   8, 107
Mixed-integer programming, *see* Mathematical programming

Minimizing total transportation time   58–59, 69–73, 91, 93, 99, 101
Modelling and solution approaches:
– hierarchical
  (*see also* Hierarchical scheduling)   40, 104, 107, 161
– interactive   60
– lexicographic   71, 76, 79, 104
– monolithic   35, 107
– multi-level
  (*see also* Multilevel scheduling)   164, 167
– single-level   165, 181
– weighting   59, 70, 78
Models:
– L   71
– L'   100
– LR   70
– L*R   72
– LRS   91
– LRS'   99
– LRλ   71
– M1   47
– M2   47
– M3   48
– M4   49
– M5   50
– M6   51
– M7   57
– M8   58
– M9   58
– M10   59
– MLS   155
– R   73
– RS   93
– RS'   101
Multilevel assembly line   149, 155
Multilevel scheduling   156–157, 159

Objective function   47–48, 50–51, 57, 70, 84, 92, 96, 104, 112, 115, 144, 148, 150
Off-line scheduling   37, 163
On-line scheduling   37–38, 163

Performing mode   108
Placement machine   10, 13, 40, 107
Precedence constraints   31, 45, 49, 58, 110
Precedence relations   44–45, 47, 53, 56, 59, 69, 87, 101, 165
– – graph of   30, 77, 81, 177, 188
Printed circuit board (PCB) assembly   8–9, 39, 107, 113

Priority index   166
Product assignments   69, 75, 78–80, 83, 92–93, 95, 104
Production schedule   160

Quadratic integer programming, *see* Mathematical programming

Real-time scheduling   37, 163
Resource   107, 163
– allocation   17, 36, 110
– requirements   109
Robot assembly cell   107

Schedule length   112, 178–179, 181, 189
Scheduling algorithm   121, 133, 156–157, 159, 176, 185
Scheduling horizon   120, 152, 159, 163–164, 169
Scheduling objectives   38
Scheduling of assembly operations   166
Scheduling of transportation operations   166
Separation problem   74–75
Surface Mount Technology (SMT)   9, 10
System locking   37, 187

Task assignments   36, 42, 58, 62–67, 72, 82–84, 93–94, 101–102, 104, 111, 164
Time-table   38, 194–195
Travelling Salesman Problem   114

Vehicle assignment   169–170, 172–173, 176, 187
Vehicle schedule   164–165, 179, 182

Working space   3–4, 19, 46, 56, 61, 81, 95, 102, 178
Workload balancing   42, 113

Druck: Strauss Offsetdruck, Mörlenbach
Verarbeitung: Schäffer, Grünstadt